Kaio Gráculo Vieira Garcia
Vânia Felipe Freire Gomes
Paulo Furtado Mendes Filho

Interaction between arbuscular mycorrhiza and manganese in sabiá seedlings

Kaio Gráculo Vieira Garcia
Vânia Felipe Freire Gomes
Paulo Furtado Mendes Filho

Interaction between arbuscular mycorrhiza and manganese in sabiá seedlings

Development and mycorrhizal colonization in soils degraded by manganese mining

ScienciaScripts

Imprint
Any brand names and product names mentioned in this book are subject to trademark, brand or patent protection and are trademarks or registered trademarks of their respective holders. The use of brand names, product names, common names, trade names, product descriptions etc. even without a particular marking in this work is in no way to be construed to mean that such names may be regarded as unrestricted in respect of trademark and brand protection legislation and could thus be used by anyone.

Cover image: www.ingimage.com

This book is a translation from the original published under ISBN 978-3-330-75549-9.

Publisher:
Sciencia Scripts
is a trademark of
Dodo Books Indian Ocean Ltd. and OmniScriptum S.R.L publishing group

120 High Road, East Finchley, London, N2 9ED, United Kingdom
Str. Armeneasca 28/1, office 1, Chisinau MD-2012, Republic of Moldova, Europe
Printed at: see last page
ISBN: 978-620-8-27126-8

CONTENTS

1 INTRODUCTION

Mineral exploration is an activity that is linked to global economic development, generating raw materials for various industries and consumer goods. Among the various minerals exploited in the world, manganese stands out mainly for its use in the manufacture of steel, as an alloying element, and in the battery production market.

Although this practice is strongly linked to social development, mining is considered to be one of the major causes of environmental pollution. In areas where mineral exploration takes place, there is intense soil movement during the opening up of sites for ore extraction, as well as the removal of natural vegetation, which can cause serious environmental consequences. Also in this context, during the mineral extraction phase, large quantities of tailings are produced, which are considered to be one of the main anthropogenic actions that cause soil pollution with metals, including manganese, which when in excess can cause toxicity, inhibit plant growth and cause changes in plant communities.

Various strategies have been developed to remediate areas contaminated with heavy metals. However, the vast majority of these involve high levels of investment and can lead to irreversible changes in soil properties and disruption of the native microbiota. In this sense, it is necessary to look for alternatives that minimize this type of impact.

A recent and very promising technique that has been used to rehabilitate these areas is phytoremediation, which basically consists of using plants to absorb and accumulate heavy metals. In this context, the association of these plants with arbuscular mycorrhizal fungi (AMF) is of great importance because, as well as providing better absorption of water and nutrients for the plants and helping plant growth in adverse locations, it can also contribute to reducing the availability of heavy metals by immobilizing them in intra- and extraradicular symbiotic structures.

However, studies involving tree legumes, more specifically Sabià (*Mimosa Caesalpiniaefolia* Benth.) associated with arbuscular mycorrhizal fungi and their effect on plant tolerance in soils impacted by manganese mining activities are still scarce.

The aim of this study was to evaluate the effects of mycorrhizal colonization on the development of Sabià (*Mimosa Caesalpiniaefolia* Benth.) in soils impacted by manganese mining and subjected to two conditions (sterile and natural).

2. HYPOTHESIS

The AMF associated with Sabiâ (*Mimosa Caesalpiniaefolia* Benth.) increase the tolerance of this plant to Mn, making it a potential phytoremediation plant.

3. LITERATURE REVIEW

3.1. General aspects of manganese (Mn) mining

Manganese mining is an activity that has made a name for itself in the Brazilian economy. A survey carried out in 2012 shows that the national production of manganese ore reached 1.1 million tons, placing Brazil fifth in the world ranking of producing countries, with 6.6% of the total produced in the world. In terms of world production, South Africa ranks first, followed by Australia, China and Gabon (DNPM, 2013).

In Brazil, the states with the highest production are Para, Mato Grosso do Sul, Minas Gerais and Bahia. Its production is led by Vale Manganês, but other small mines also exploit the ore (CASTRO, 2011). According to Vidal *et al.* (2005), manganese production in the state of Ceará is basically concentrated in the Greater Fortaleza region, and is characterized by small deposits, of which no more than 10 million tons (IBRAM, 2010). Information from the Brazilian Mining Institute (IBRAM, 2010) indicates that the number of requests for prospecting in areas with manganese ore in Ceará reached 59 applications, of which 44 were authorizations and 1 was a mining application in the municipality of Ocara - CE.

Currently, a large part of the total production of manganese ore is consumed in the manufacture of steel, as an alloying element, due to its desulphurization property. In the non-metallurgical industry, this element is also used in the manufacture of electrolytic batteries, fertilizers, ceramics, paints, varnishes and chemical reagents. In addition to these, there is also another little-known market for this ore, which is for some vitamins (GONÇALVES; SERFATY, 1976; SANTANA, 2009). The beneficiation process for this ore generally consists of separating the organic material, crushing, classifying the particle size and washing to remove the fine fraction.

Despite its importance to the country's economy, manganese mining can cause numerous environmental and social impacts. Open-cast mines cause soil degradation and changes due to the removal of vegetation. The dust released by the mining industries from the processing process can compromise the quality of the soil and cause serious respiratory problems in humans through permanent contact. According to Cleolin (2010), prolonged exposure to high levels of manganese can cause chronic poisoning, known as "manganism". Who (1999) reports that in some cases the symptoms of "manganism" are similar to Parkinson's disease.

Some studies have been carried out in an attempt to solve the problem of recovering areas degraded by mining. Castro (2011) carried out studies on the use of manganese processing tailings by the ceramics industry, with the aim of evaluating the possibility of using these tailings as a raw material. Wiltshire (2002) in a literature review investigates the potential applications for manganese tailings, highlighting the use of tailings for agricultural applications, as well as their use as a raw material for glass, plastics and rubber additives.According to D'Agostino (2008) tailings are materials originating from mineral processing processes, basically made up of material of no economic interest.

Although some studies have been carried out with the aim of using tailings as a raw material, large quantities are still released and deposited in the soil. This raises the need for new research aimed at using this material as a substrate for re-establishing vegetation.

3.2. Manganese in the soil

Manganese is a chemical element that belongs to the class of transition metals and has an atomic number of 25, placing it in group 7 of the periodic table. This metal has a light gray color and is found in more than a hundred minerals (JACOBI, 2014). One of the forms of this mineral most commonly found in the soil are manganese oxides, including pyrolusite (MnO_2), manganite (MnOOH) and haussmannite (Mn_3O_4). In addition, this element is considered to be one of the most abundant transition metals after iron and titanium, comprising approximately 0.1% of the earth's crust (RAMOS, 2010).

With regard to manganese cycles in the soil, the bivalent and trivalent forms of this metal are involved, and there is a dynamic equilibrium between them. The bivalent form is transformed by biological oxidation into the trivalent form, which is then reduced to Mn^{++} in very acidic soils. In alkaline soils, the divalent form practically disappears. The oxidation and reduction potential of this element is directly related to the activity of micro-organisms that can alter the pH (WHO, 1981; MARTINS, 2003).

In the soil, manganese can be found in three oxidation levels: Mn^{2+} (form absorbable by plants), Mn^{+3} and Mn^{+4} (less mobile). According to Marschner (1995), divalent or reduced manganese (Mn^{2+}) is generally the predominant form in the soil solution, depending on pH, humidity and aeration conditions.

In Brazilian soils, the total manganese content ranges from 10 to 4000 mg kg^{-1} , while its soluble content varies from 0.1 to 100 mg kg^{-1} (MALAVOLTA, 1980).CONAMA currently establishes soil contamination values for a large number of metals. However, there are still no reference or intervention values available for manganese in the soil. According to Rufino (2006), a contaminated area is classified according to its impact on vegetation or in comparison with adjacent soil of the same classification, as well as the impact on groundwater or surface water.

3.3. Manganese in the plant

Manganese is an element of great importance for plant metabolism. In addition to acting in the synthesis of chlorophyll and photosynthesis, it participates in the activation of different enzymes and the functioning of chloroplasts (BUCKMAN; BRADY, 1974; EPSTEIN, 1975; MALAVOLTA, 1977; TAIZ; ZEIGER, 2004).

With regard to the absorption of this element by plants, the greatest proportion of Mn contact from the soil solution to the root is by diffusion and interception, except in very rich soils where the phenomenon that plays the greatest role is mass flow (MALAVOLTA, 2006).

Manganese is absorbed by plants in the form of Mn^2 +, being transported from the roots to the aerial part via the xylem, following the transpiration current, undergoing little remobilization. Three compartments of Mn in the roots are recognized. The first refers to the exchangeable fraction, in the apoplast, where it remains adsorbed to the negative charges of the constituents of the cell wall. The second, called labile, is the Mn found in the cytoplasm, while the last, non-labile, refers to the Mn deposited in the vacuoles (MUKHOPADHYAY; SHARMA, 1991).

Manganese levels in plants can vary between 5 and 1500 mg kg^{-1} in the dry matter, depending on the species and the part to be analyzed. The values considered adequate for normal growth and development are between 20 and 500 mg kg^{-1} , where below this can cause deficiency and above, toxicity (DECHEN; NACHTIGALL, 2006).

Generally, the symptoms of Mn toxicity in plants are characterized by a decrease in growth rate (MORONI *et al*., 2003), brown necrotic spots and inter-veinal chlorosis on the leaves (WISSEMEIER; HORST, 1991). In addition, Mn toxicity can be intensified when other available elements such as Ca, Mg, K, Fe and Si are in small quantities (ABOU *et al.,* 2002), possibly because they do not act antagonistically to Mn.

Mn toxicity in plants has been reported by different authors. Rufino (2006) observed symptoms of manganese toxicity in two species of the *Brassicaceae* family from a dose of 200 mg kg-1, showing damage such as necrosis at the tips of the leaves, curling of the leaf edges, loss of green color and reduced growth. On the other hand, Xue et *al.* (2004), studying *Phytolacca acinosa* Roxb. as a manganese hyperaccumulator plant, reported toxicity symptoms when the Mn concentration was above 8000 µmol L-1. In some varieties of forage grasses, toxicity symptoms are observed in the range between 60 and 217mg dm^{-3} (NARDIS, 2012). Javan and Bingham (1981), who studied Mn toxicity in coffee seedlings (*Coffea arabica* L.) cv. Catuai vermelho, with the aim of characterizing and determining Mn toxicity, observed Mn toxicity symptoms when the leaf content was 1200 µg g^{-1} of Mn.

3.4. Phytoremediation in degraded areas

The contamination of soil by heavy metals due to mining activities is currently considered to be a major environmental concern. Several problems are generated by mining, among which the best known are soil and water contamination, which have only received due importance since the twentieth century when serious human health and environmental problems were publicized by the world media (BRANCHER;

RODRIGUES, 2010).

Recovering areas degraded by mining activities is considered a challenging ecological problem (MUKTA; SREEVALLI, 2010). Generally, in mining tailings areas, the soils are characterized by being poorly structured, having poor availability of essential nutrients, as well as containing potentially toxic levels of various metals.

In this sense, many techniques have been developed and tested in order to mitigate the damage caused to nature. However, conventional techniques (excavation, landfill and physical-chemical treatment) involve high investment, as well as generating irreversible changes in soil properties and disturbing the native microbiota. In view of this, the search for new, economically sustainable techniques has increased in recent years.

Phytoremediation is a technique that uses plants to extract, degrade, contain or immobilize contaminants in the soil (VASCONCELOS *et al.*, 2012). Several studies have highlighted the potential of this technique in the remediation of contaminated soils (CAIRES *et al.*, 2011; MAGALHAES *et al.*, 2011; LOTFY; MOSTAFA, 2014). Its main advantages are related to the low cost of investment and operation, *in situ* applicability, and minimal generation of degradation and destabilization of the area to be decontaminated (CHAVES *et al.*, 2010).

Among the various phytoremediation techniques, phytostabilization consists of immobilizing the contaminating metal in the soil through absorption and accumulation inside or outside the root or precipitation of the metal in the root zone (BRANCHER; ROGRIGUES, 2010), thus reducing leaching, mobility and the entry of metals into the food chain (Figure 1).

Figure 1 - Phytostabilization scheme. Source: BRANCHER; ROGRIGUES (2010).

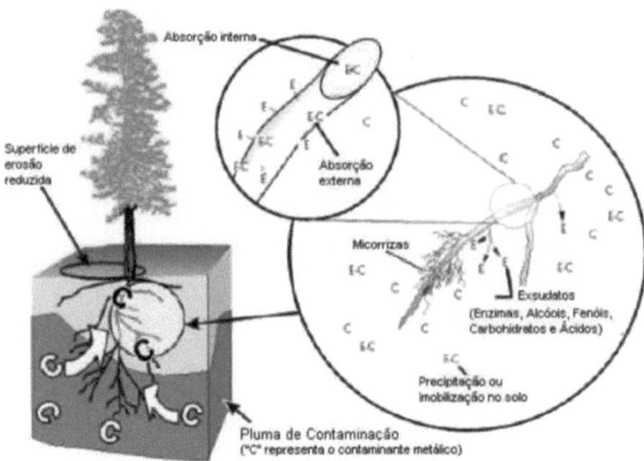

In addition, the plants used in this technique can also help alter the chemical form of contaminants through changes in the environment around their roots, such as pH and redox potential (VASSILEV *et al.*, 2004).

According to Lee *et al.* (2014), in order to successfully recover degraded areas using the phytostabilization technique, it is necessary to select suitable plant species that have characteristics such as a deep root system and a large amount of biomass in the presence of high metal concentrations.

Several authors (SOUZA *et al.*, 2012; ARAÙJO; COSTA, 2013; NASCIMENTO *et al.*, 2014) point to the use of leguminous plant species as promising in the recovery of degraded areas. Studies carried out in Brazil by Souza *et al.* (2010) using some species of leguminous trees, show the potential of the sabia species (*Mimosa Caesalpiniaefolia*) as one of the most promising in the process of phytostabilization. According to them, this species was one of the most tolerant in the seedling phase in areas contaminated by lead. Araujo and Costa (2013) when working with some species of leguminous trees aimed at recovering areas degraded by iron mining, observed that the species *Mimosa caesalpiniaefolia* showed a greater accumulation of manganese (139.60 μg $root^{-1}$ $plant^{-1}$) in the root system than the other species studied. Nascimento *et al.* (2014), when researching the development of knew seedlings in degraded soils subjected to organic fertilization, found that this species showed greater efficiency in the absorption of all the metal micronutrients studied (Mn, Cu, Zn and Fe) compared to the other species studied.

Special attention has also been paid to the association of plants with arbuscular mycorrhizal fungi (AMF) in the process of phytostabilization, given that these fungi have some mechanisms that can increase the absorption of macro and micronutrients, including metals, into the host root (MARSCHNER, 1998), thus increasing the tolerance of plants to phytotoxicity and helping in the revegetation process (LELIE *et al.,* 1999).

3.5. Sabià (*Mimosa Caesalpiniaefolia* Bent.)

The species *Mimosa Caesalpiniaefolia* Benth. commonly known as Sabià in the northeast of Brazil, is also known as sansao-do-campo in the states of Rio de Janeiro and Sao Paulo. Belonging to the Mimosaceae family, it is a species native to the Northeast region of Brazil (LACERDA *et al*., 2006), occurring in caatinga areas of Piaui, Pernambuco, Alagoas, Rio Grande do Norte, Paraiba, Bahia and Cearà (FIGUEIRÔA *et al*., 2005).

In general, it is considered a small tree, reaching a height of approximately seven to eight meters, with a trunk well endowed with aculei. Its young stem is characterized by being sparsely thorny, losing the thorns as the bark thickens (CARVALHO, 2007). The leaves of this species are compound, bipinnate and alternate, with four to six opposite leaflets. In addition, this tree has a deep root system, concentrated in the first twenty centimetres of the soil, which can reach up to six meters in length, thus favoring water absorption and helping the species to thrive in limiting environmental conditions (RIBASKI *et al*., 2003; ALENCAR, 2006).

Sabià occurs spontaneously in semi-wet areas of the Caatinga, as well as in drier areas. This species grows preferentially in deeper soils and does well in poor soils. However, when cultivated in fertile soils, by the end of the third to fourth year, they can already provide wood for the production of stakes used in the construction of fences (RIBASKI *et al*., 2003).

In the Northeast, this species stands out as one of the main sources of cuttings for the production of fences, especially in the state of Cearà. According to Leal Jùnior *et al.* (1999), this region is considered the largest producer and exporter of Sabià cuttings, with most of the production concentrated in the north of the state. Other aspects, such as the production of wood used as a source of energy, the formation of windbreaks and the production of fodder for animals in drier periods, are characteristics that make this species a tree of multiple use in Brazil (RIBASKI *et al.*, 2003).

According to Lorenzi (2000) Sabià is of great interest in the recovery of degraded areas. During the growth of this species, part of the biomass produced returns to the soil in the form of branches, leaves and reproductive structures, forming a layer known as litter. This condition plays an important role in nutrient cycling, soil stabilization, increased biological activity and attenuation of erosion processes, thus encouraging the establishment of other more demanding species and consequently the recovery of degraded areas (FRANCO *et al.*, 1992; COSTA *et al.*, 2004).

In addition, this species has the characteristic of associating symbiotically with nitrogen-fixing bacteria and arbuscular mycorrhizal fungi, benefiting its development in adverse locations, especially in environments with nutritional deficits.

Studying the double inoculation of rhizobium and mycorrhiza in sabià seedlings aimed at recovering degraded areas, Pontes *et al.* (2012) reported that this technique proved to be more efficient for maximizing the production of plants of this species, as well as inducing greater nodulation and symbiotic efficiency. Similar results were also observed by Burity *et al.* (2000) in sabià seedlings inoculated with rhizobium and AMF.

Oliveira and Alixandre (2013), when working with sabia seedlings under phosphorus levels in a yellow latosol, reported a greater increase in the growth of the aerial part and the diameter of the collar when inoculated with *Claroideoglomus etunicatum* and native AMF compared to plants in a substrate without inoculation.

When evaluating the potential of forest species for microbial management in the recovery of areas degraded by cassiterite mining, Mendes Filho *et al.* (2009) reported that the species *Mimosa Caesalpiniaefolia* had a significant influence on the height and dry mass of the aerial part, as well as the nitrogen and phosphorus contents when inoculated with AMF in the presence of organic matter compared to the control.

3.6. Arbuscular mycorrhizal fungi (AMF) and their importance

Arbuscular mycorrhizal fungi (AMF) form an important symbiosis with the vast majority of plants, occurring in almost all ecosystems, being formed in approximately 80% of terrestrial plant species with fungi from the Glomeromycota phylum (SCHUβLER *et al.*, 2001), including in soils altered by mining activities (DEL VAL *et al.*, 1999).

This type of symbiosis, known as "mycomycotic association", is an important survival mechanism for some plants in adverse environments. In addition to improving the absorption of water and nutrients for plants, arbuscular mycorrhizal fungi are involved in the uptake and detoxification of various pollutants, including heavy metals (TURNAU *et al.*, 2006).

The importance of AMF associated with plants in soil remediation has been reported by several authors (CHRISTIE *et al.*, 2004; CHEN *et al.*, 2005; SILVA *et al.*, 2006), because, as well as providing better phosphorus (P) absorption and aiding plant growth (RABIE, 2005), they can also contribute to reducing the availability of heavy metals to plants. However, the mechanisms involved in the protection of plants by AMF against metal toxicity are still unclear.

Some probable mechanisms have been reported in the literature, including the dilution effect of trace elements in plant tissues (SCHNEIDER *et al.*, 2013), chelation of metals by compounds secreted by AMF (VODNIK *et al.*, 2008) and reduced absorption of metals due to their retention and immobilization in fungal structures, increasing their retention in the roots and reducing translocation to the aerial part, reducing the likelihood of toxicity in plants (CHRISTIE *et al.*, 2004).

Several studies have been carried out using Ieguminous trees and arbuscular mycorrhizal fungi (AMF) to establish plants in different types of environments, including areas degraded by mining (LINS et al., 2007).

In *Leucaena Ieucocephala* (Lam.) de Wit. seedlings in caatinga soils under the impact of copper mining, LINS *et al.* (2007) reported better leaf formation when inoculated with AMF. MA *et al.* (2008) also working with *Leucaena leucocephala* associated with earthworms and arbuscular mycorrhizal fungi in areas of Zn and Pb tailings, observed benefits for plant growth and nutrition, as well as protection against the absorption of toxic metals.

Sugai *et al.* (2011) described better development of the aerial part of angico seedlings at 150 days after germination when they were inoculated with a combination of *Glomus etunicatum* and *Gigaspora margarita*, both in preserved and anthropized soil.

3.7. Arbuscular mycorrhizae x manganese toxicity in plants

Mycorrhizal association is considered an important mechanism which provides a number of benefits for the host plant. In addition to improving nutritional status, there are some reports that AMF can reduce the availability of some metals to plants, including manganese.

According to Nogueira and Cardoso (2002), Mn content and absorption are generally lower in mycorrhized plants. The same authors also explain that this effect can contribute to increasing plant tolerance to excess Mn. The mechanisms by which AMF provide protection to plants against an excess of this element in the soil are still unclear. However, Nogueira and Cardoso (2002) attribute the protection of plants against excess Mn to an indirect effect of mycorrhizal fungi which can cause changes such as alterations in the pattern of root exudation and consequently in the composition of the communities of Mn-oxidizing and Mn-reducing microorganisms in the rhizosphere and, consequently, in plant nutrition in relation to Mn.

Cardoso *et al.* (2003), studying the absorption and translocation of manganese by mycorrhized soybean plants under increasing doses of this nutrient, observed that the AMF *Glomus macrocarpum* promoted the greatest protection against excess Mn, resulting in greater retention of this element in the roots and less translocation to the aerial part. While Nogueira and Cardoso (2002), when researching the interaction between arbuscular mycorrhizal fungi and the soil microbial community in mitigating

Mn toxicity in soybeans, observed that the mycorrhized plants showed greater growth and mitigation of Mn toxicity, especially when the native microbial community was reintroduced. In a more recent study, Becerril *et al.* (2013) observed that the colonization of *Ambrosia psilostachya* plants by AMF reduced Mn absorption in the plant tissues of this species in contrast to non-mycorrhized plants.

4. MATERIAL AND METHODS

4.1. Location and Climate of the Experimental Area

The experiment was conducted in a greenhouse at the Department of Soil Sciences (DCS) of the Agricultural Sciences Center of the Federal University of Cearà (UFC), located on the Pici Campus, Fortaleza, Cearà (3°45'47' south latitude and 38°31' 23' west longitude, average altitude 47m), for a period of 60 days (December 2014 to February 2015). The region's climate is classified as tropical, hot with an average annual temperature and rainfall of 27°C and 1600mm respectively, and is characterized by Koeppen as type Aw'.

4.2. Soil

The soils used in this experiment were collected at a depth of 0-20 cm in a manganese mineral exploration area located in the municipality of Ocara-CE, 101 km from Fortaleza. The soils were collected from two different areas (Area - 1) and (Area - 2) (Figure 2). The first is an area of preserved forest, free from the mining process (A1) and the second is an area degraded by Mn mining tailings (A2). After collection, the soil was sieved through a 4 mm mesh sieve, and part of both soils was sterilized in an autoclave at 121 °C at 1 atm pressure for two hours and incubated for 14 days, while the other non-autoclaved part remained in natural conditions.

Figure 2- Location of the study area. Source: Google Earth. 2015.

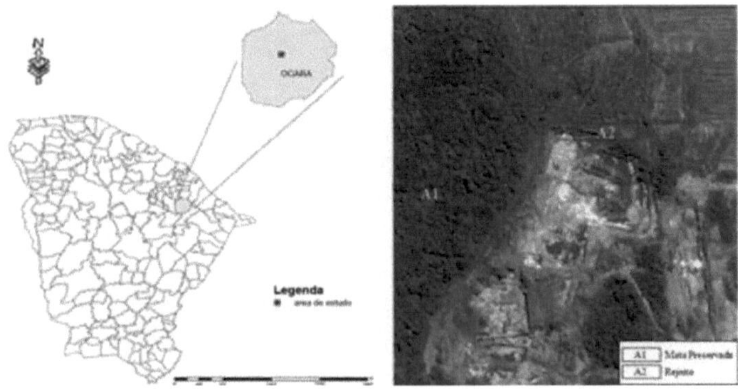

4.2.1. Chemical and Physical Soil Characterization

The soil used in this experiment had the following physical and chemical characteristics (Table 1).

Table 1 - Chemical and physical characteristics of the soil from the preserved forest (SMP) and the soil degraded by the

(SDR) used in the experiment before (A) and after (D) 14 days of autoclaving.

Sample		pH	Al	Ca	Mg	In	K	S	H+Al	P	N	M.O.
		(H_2 O)			------ (cmolc/kg) -----				------	(mg/kg)	(g/kg)	
SMP	A	6,70	0,10	3,7	1,1	0,04	0,35	5,19	2,20	6,8	1,12	15,52
	D	7,57	0,10	3,4	1,4	0,10	0,34	5,23	0,60	10,0	1,12	
SDR	A	4,95	0,20	1,4	1,2	0,12	0,15	2,87	4,00	2,9	0,28	3,93
	D	5,74	0,10	1,7	1,6	0,24	0,18	3,72	4,00	4,0	0,28	

Sample		Mn	Fe	Cu	Zn	Sand	Silt	Clay	Classification
			(mg/kg)		------	(%)	(%)	(%)	Textural
SMP	A	253,57	13,72	2,57	6,26	76,97	13,15	9,88	Sandy loam
	D	520,17	42,1	3,26	8,17				
SDR	A	425,80	82,81	2,39	2,91	72,81	16,62	10,57	Sandy loam
	D	746,71	137,99	4,32	4,61				

N (micro-Kjeldahl method); / P (nitroperchloric digestion, colorimetry); / K, Na (nitroperchloric digestion, flame spectrophotometry);/ Ca, Mg, Cu, Mn, Zn and Fe (nitroperchloric digestion, atomic absorption spectrophotometry);/ Particle size analysis (pipette method - Embrapa, 1997).

4.3. Setting up and conducting the experiment

The soils were distributed in plastic pots with a capacity of 5 liters of soil, using 4 kg of soil per pot, which received a basic fertilizer with: 27 mg of N, 100 mg of K, 40 mg of P and 30 mg of Ca per pot. The nutrient sources used were: $NH2CONH2$, KCl, $Ca(H2PO4)2.H2O$ and CaSO4 respectively.

The Sabià seeds (*Mimosa CaesalpiniaefoliaBenth.*) were purchased from the company Biosementes, located at Km. 24, Rodovia Ilhéus/Itabûna, BR 415.

The seedlings were grown in Styrofoam trays with 128 cells, placing two seeds per cell at a depth of 2 cm. The substrate used to produce the seedlings in the trays was autoclaved sand. At 11 DAS (days after sowing), thinning was carried out in each cell, selecting the most vigorous plant.

After germination, the seedlings were inoculated with 40g of inoculum-soil containing spores and fragments of maize roots (*Zea mays* L.) colonized by the species *Glomus clarum* and *Glomus etunicatumisolated* and in mixture, when the seedlings were transplanted into plastic pots containing 4 kg of soil according to each treatment proposed in the experiment. Each pot received two seedlings (Figure 3).

The isolates used in this experiment were obtained from the collection of AMF cultures at the Soil Microbiology laboratory of the Federal University of Ceará (UFC), Pici campus.

The inoculum was placed approximately 4 cm below the surface of the soil in the container. In order to re-establish the microbial community in both types of soil, except for the AMF, 8 mL of a "filtrate" from the soil of the preserved forest area (A1) which had not been autoclaved was added to all the pots in this treatment and 8 mL of a "filtrate" from the soil of the area degraded by the tailings (A2) which had not been autoclaved was added to all the pots in this type of soil. To do this, the "filtrate" was obtained by stirring 10 cm^3 of soil from each area separately in 1 dm^3 of distilled water, under vigorous stirring. The suspension was then passed through a series of sieves with decreasing mesh sizes, the last one being 44 μm, in order to retain the AMF propagules, whose absence in the effluent was confirmed under a stereoscopic microscope.

The plants were kept in a greenhouse and irrigated daily throughout the experiment, which lasted 60 days.

Figure 3- (A) Preparing the soil for transplanting and (B) 1 DAT (day after transplanting) of the sabià seedlings (*Mimosa caesalpiniaefolia* Benth.).

4.3.1. Variables evaluated

They were evaluated 60 days after transplanting:

I) Plant height: determined from the surface of the substrate to the insertion of the last leaf using a millimeter ruler (Figure 4);

II) Stem diameter: measured at 3 cm from the ground using a digital caliper;

III) Number of leaflets: by direct counting;

IV) relative chlorophyll index (CRI)

The relative chlorophyll index was estimated using the portable chlorophyll meter Soil-Plant Analysis Development - SPAD-502 (Minolta Corp., Japan), between 8:00 and 10 am. The index refers to the average of the four terminal leaflets.

Figure 4 - Determination of height (A), neck diameter (B), number of leaflets (C) and relative chlorophyll index (D) of sabia plants (*Mimosa Caesalpiniaefolia* Benth.).

V) Dry matter mass of root and aerial part:

To measure the masses of the dry matter, the plant material was dried in an oven with forced air circulation at 65 °C until it reached a constant mass. The material was then weighed on an analytical balance to the nearest 0.0001g.

VI) Root system length:

Determined from the neck of the plant to the end of the main root, using a millimetre ruler.

VII) Accumulation of macronutrients (N, P and K) in the aerial part and manganese (Mn) in the aerial part and root:

After drying in an oven, the plant material was ground for chemical analysis. The extracts for the analysis of phosphorus (P), potassium (K) and manganese (Mn) were obtained by nitroperchloric digestion following the procedures described by Malavolta *et al.* (1997). Mn levels were determined by atomic absorption spectrophotometry, K by flame photometry and P by colorimetry.

Total nitrogen levels were determined using the semi-micro Kjeldahl method, adapted from Bremner and Mulvaney (1982) and Tedesco et *al.*, (1995), which is based on distillation and vapor drag.

VIII) Mn translocation factor:

The translocation factor (FT) was obtained according to Santillanet *al.* (2010) and calculated using the following formula:

Mn concentration in the aerial part

Mn concentration in the root

IX) Mycorrhizal efficiency:

Mycorrhizal efficiency was determined using the formula ME = [(total biomass in plants inoculated with AMF - total biomass in plants not inoculated with AMF) / total biomass in plants not inoculated with AMF] x 100 (CHAVES *et al.*, 2005).

X) Mycorrhizal root colonization

To assess root mycorrhizal colonization, the roots were washed in running water and placed in a container with a 70% alcohol solution. The roots were clarified and stained for colonization analysis according to the methodology proposed by Phillips and Hayman (1970) and adapted by Koske and Gemma, (1989) and Grace and Stribley, (1991). The percentage of colonization was obtained according to McGONIGLE *et al.*,

(1990).

XI) AMF spore density in the soil.

The density of AMF spores was determined by extracting 100g of soil via wet sieving from each treatment sample, following the procedures described by Gerdemann and Nicholson (1963).

4.4. Experimental Design

The experimental design adopted was entirely randomized, in a 2x2x4 factorial arrangement, considering: i) two types of soil (degraded by tailings and preserved forest); ii) two soil conditions (sterile and natural); iii) four inoculation treatments (non-inoculated (control), inoculated with *Glomus clarum*; inoculated with *Glomus etunicatum* and inoculated with *Glomus clarum* + *Glomus etunicatum*(Mix)), with four replications, constituting 64 experimental units (Figure 5). The mycorrhizal efficiency variable was analyzed as a 2x2x3 factorial, since the control (non-inoculated) was not considered in the analysis.

Figure 5 - Schematic diagram of the experiment. Source: GARCIA, K. G. V. 2015.

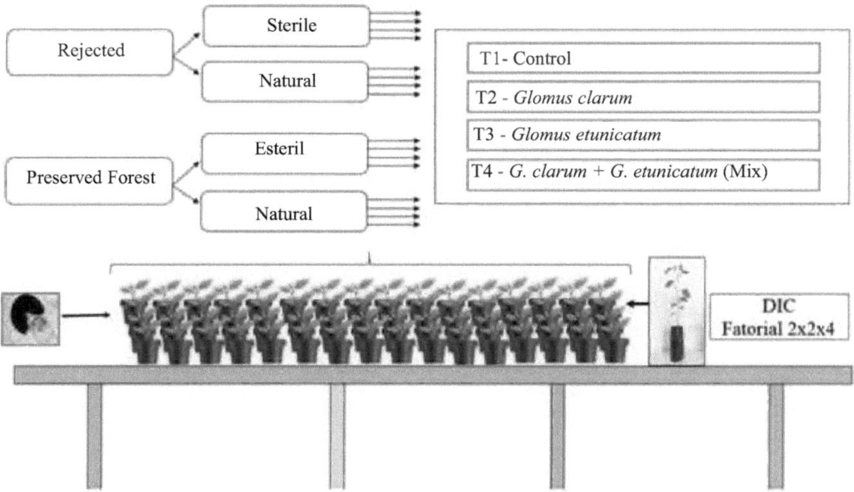

4.5. Statistical Analysis

The data were subjected to analysis of variance and the means of the treatments were compared using the Scott-Knott test at 5% probability, using the ASSISTAT statistical program version 7.7 Beta (SILVA, 2013).

5. RESULTS AND DISCUSSION

5.1. Growth variables

According to the analysis of variance (Table 2) it can be seen that the growth variables were significantly influenced by the factors soil condition (sterile and natural) and inoculation treatment with AMF, but only height (ALT), number of leaflets (NF), neck diameter (DC), mass of aerial matter (MMSPA) and root mass (MMSR) were influenced by the factor soil type (degraded by tailings and preserved forest). The interaction between soil type and soil condition was only significant for neck diameter (CD) and root system length (RSC). In relation to the interaction between soil type and inoculation treatment, only the collar diameter (CD) and root system length (RSC) variables were not significantly influenced.

Table 2 - Summary of the analysis of variance for height (ALT), number of leaflets (NF), neck diameter (DC), root system length (CSR), relative chlorophyll index (IRC), mass of dry matter of the aerial part (MMSPA) and root (MMSR) for the treatment factors and their interactions in *Mimosa caesalpiniaefolia* Benth.

F. V.	G.L.	MEAN SQUARE						
		ALT	NF	DC	CSR	IRC	MMSPA	MMSR
TS (A)	1	$19{,}88^{**}$	$41{,}77^{**}$	$4{,}33^{*}$	$0{,}58^{ns}$	$1{,}54^{ns}$	$79{,}05^{**}$	$4{,}84^{*}$
CS (B)	1	$366{,}92^{**}$	$273{,}34^{**}$	$160{,}33^{**}$	$12{,}49^{**}$	$172{,}41^{**}$	$236{,}87^{**}$	$144{,}62^{**}$
AMF (C)	3	$78{,}95^{**}$	$54{,}97^{**}$	$56{,}31^{**}$	$3{,}71^{*}$	$43{,}68^{**}$	$114{,}32^{**}$	$37{,}64^{**}$
Int. (AxB)	1	$2{,}92^{ns}$	$0{,}09^{ns}$	$7{,}89^{**}$	$7{,}80^{**}$	$1{,}27^{ns}$	$2{,}01^{ns}$	$0{,}03^{ns}$
Int. (AxC)	3	$7{,}01^{**}$	$15{,}89^{**}$	$2{,}54^{ns}$	$2{,}45^{ns}$	$3{,}68^{*}$	$16{,}54^{**}$	$3{,}64^{*}$
Int. (BxC)	3	$72{,}02^{**}$	$37{,}16^{**}$	$26{,}14^{**}$	$4{,}09^{*}$	$11{,}92^{**}$	$60{,}12^{**}$	$16{,}46^{**}$
Int. (AxBxC)	3	$2{,}83^{*}$	$0{,}78^{ns}$	$2{,}70^{ns}$	$7{,}71^{**}$	$3{,}04^{*}$	$5{,}97^{**}$	$3{,}29^{*}$
Total	63	-	-	-	-	-	-	-
C.V.%	-	10,81	14,17	13,57	13,35	7,71	16,06	26,44

G.L - degree of freedom; (A) - soil type; (B) - soil condition; (C) - AMF inoculation treatment; CV - coefficient of variation; **, *, ns - significant by the Scott-Knott test at 1%, 5% and non-significant, respectively.

The interaction between soil condition and inoculation treatment (Table 2) showed a significant influence on all the variables analyzed. When evaluating the interaction between the three factors (Table 2), it was found that all the variables underwent significant changes, with the exception of the number of leaflets (NF) and neck diameter (DC). In order to better visualize this result, the effect of the inoculation treatments on each type of soil was split: Degraded by tailings and Preserved forest.

5.1.1. Height

In the height variable (Table 3), it can be seen that plants grown in soil degraded by Mn mining tailings and preserved forest only showed a significant influence between AMF inoculation treatments in the sterile soil condition. In the soil degraded by the tailings, the Mix and *G. etunicatum* treatments provided increases of 71.5% and 65.9%, respectively, compared to the absence of inoculation. When using this same comparison for the plants grown in the preserved forest soil, it can be seen that the treatments with *G. etunicatum* followed by Mix promoted increases of 78.3% and 75.6%, respectively, while the treatment with *G. clarum* showed similar results to the control. In general, when analyzing each inoculation treatment between the soil conditions (barren and natural), it can be seen that plants grown under natural conditions showed greater height values when compared to the barren condition, with the exception of the Mix treatment in the soil degraded by the tailings and the treatment with *G. etunicatum* in the preserved forest.

Table 3 - Height (ALT) of sabia plants (*Mimosa Caesalpiniaefolia* Benth.) subjected to two types of soil (degraded by tailings and preserved forest), two conditions (sterile and natural) and four inoculation treatments with AMF.

Inoculation treatments (AMF)	Degraded by Tailings		Preserved Forest	
	Sterile	Natural	Sterile	Natural
		ALT (cm)		
Control	11.37 cB	40.37 aA	10.25 bB	46.62 aA
Glomus clarum	15.25 cB	44.50 aA	10.81 bB	46.18 aA
Glomus etunicatum	33.43 bB	40.50 aA	47.37 aA	48.06 aA
Mix	39.93 aA	41.62 aA	42.12 aB	49.81 aA

[1] - Averages followed by the same lowercase letter in the column for each soil condition and uppercase letters in the row for each soil type separately do not differ by the Scott-Knott test at 5% significance.

These results partially agree with those obtained by Andrade *et al.* (2009) when they studied the growth and absorption of pig bean plants inoculated with arbuscular mycorrhizal fungi in a soil with increasing amounts of zinc. According to the authors, the mycorrhized plants showed greater growth compared to the non-mycorrhized plants. A similar result was also observed by Nogueira *et al.* (2007) when working with soybean plants inoculated with *G. macrocarpum* and *G. etunicatum* grown in a soil with a high manganese content.

It is also important to note that during the entire period of the experiment, the control plants grown in the soil degraded by the tailings, especially in the sterile condition, showed symptoms of toxicity (Figure 6), which may have inhibited their development compared to the inoculated plants. Cardoso *et al.* (2003) point out that the reduction in Mn toxicity in plants inoculated with arbuscular mycorrhizal fungi can be attributed to an indirect effect of the higher P content in the tissues, through the formation of a complex with Mn inside the plant which reduces its activity. Although the preserved forest soil also had considerable levels of manganese, the appearance of toxicity symptoms in the control plants (tailings) was less aggressive compared to the plants grown in the soil degraded by the tailings. The greater expressiveness of the toxicity in these treatments may be related to sterilization, since after carrying out this procedure, an increase in manganese levels was observed in the chemical analysis. This increase was around 42.9% for the soil degraded by the tailings and 51.2% for the preserved forest soil (Table 1). According to Nogueira *et al.* (2003) the increase in manganese availability after autoclaving is attributed to the thermal breakdown of organic chelating agents, one of the main factors regulating Mn availability.

Figure 6 - Toxicity symptoms at 15 (A), 30 (B), 45 (C) and 60 days after transplanting (D) in sabià seedlings (*Mimosa Caesalpiniaefolia* Benth.) from the sterilized control treatment grown in soil degraded by tailings.

5.1.2. Number of leaflets

Based on the average values of the soil type x inoculation treatment interaction (Table 4), it was found that the plants grown in the preserved forest soil had the highest number of leaflets when inoculated with *G. etuniCatum* (125.18 und. plant^{-1}).

Table 4 - Number of leaflets (NF) in sabia plants (*Mimosa Caesalpiniaefolia* Benth.) subjected to two types of soil (degraded by tailings and preserved forest) and four treatments of inoculation with AMF.

Soil type	Inoculation treatment			
	Control	*G. clarum*	*G. etunicatum*	Mix
		NF (und. plant)$^{-1}$		
Degraded by Tailings	56.25 aB	66.43 aB	75.81 bA	86.87 bA
Preserved Forest	63.56 aC	63.18 aC	125.18 aA	107.25 aB

[1-]Averages followed by equal letters in the row and in the columns do not differ by the Scott-Knott test at a significance level of 1%.

Analyzing the influence of the inoculation treatments on each type of soil, it can be seen that the plants grown in the soil degraded by the tailings suffered the influence of inoculation with Mix and *G. etunicatum* in relation to the absence of inoculation, promoting an increase in the number of leaflets of 34.4% and 25.8%, respectively, in relation to the control. In the preserved forest soil, statistical differences were also observed between the inoculation treatments, with *G. etunicatum* showing the highest percentage increase in the number of leaflets of the sabià plants (49.2%) compared to the control.

An increase in leaf area in plants colonized by arbuscular mycorrhizal fungi was also observed by Andrade *et al.* (2009) in pig bean plants grown in soil contaminated with zinc and by Souza *et al.* (2011) in *Calopogonium mucunoides* plants in soil contaminated with lead.

In the soil condition x AMF inoculation treatment interaction (Table 5), it can be seen that plants grown in soil under natural conditions had the highest values for the number of leaflets, especially when inoculated with *G. etunicatum* (110.43 und. plant^{-1}).

Table 5 - Number of leaflets (NF) in sabia plants (*Mimosa caesalpiniaefolia* Benth.) subjected to two soil conditions (sterile and natural) and four inoculation treatments with AMF.

Soil Condition	Inoculation treatment			
	Control	*G. clarum*	*G. etunicatum*	Mix

		NF (und. plant)[-1]		
Sterile	19.81 bB	27.68 bB	90.56 bA	89.81 bA
Natural	100 aA	101.93 aA	110.43 aA	104.31 aA

[1]-Averages followed by equal letters in the row and in the columns do not differ by the Scott-Knott test at a significance level of 1%.

However, when analyzing each condition (sterile and natural), a statistical difference was only observed in the sterile condition, in which the *G. etunicatum and* Mix treatments showed better results, positively affecting the number of leaflets by 78.1% and 77.9% compared to the control, while the results for *G. clarum* were similar to the absence of inoculation.

Despite the higher number of leaflets observed in the natural condition, the inoculation treatments did not influence the increase in this variable under this condition.

Possibly, the absence of a significant effect in the inoculation treatments in the natural condition was due to the greater competitiveness of the native AMF compared to the inoculated AMF. The higher values for the number of leaflets produced in this condition could probably be related to a lower concentration of Mn and a greater adaptability and diversity of native AMF present in this condition compared to the barren one. These results partially agree with those observed by Costa *et al.* (2005) in mangabeira seedlings. According to the authors, seedlings inoculated with AMF showed higher leaf area values under natural conditions compared to those grown in disinfected soil, with the exception of the AMF *Gigaspora albida.*

5.1.3. Neck diameter

Considering the soil type x soil condition interaction (Table 6), the highest values for neck diameter were observed in sabia plants grown in soil under natural conditions for both types of soil analyzed. These values were 4.13 mm for soil degraded by tailings and 4.74 mm for preserved forest.

Table 6 - Collar diameter (CD) of sabia plants (*Mimosa Caesalpiniaefolia* Benth.) subjected to two types of soil (degraded by tailings and preserved forest) and two conditions (barren and natural).

Soil type	Soil Condition	
	Sterile	Natural
	DC (mm)	
Degraded by Tailings	2.91 aB	4.13 bA

Preserved Forest	2.82 aB	4.74 aA

[1]-Averages followed by equal letters in the row and in the columns do not differ by the Scott-Knott test, at a significance level of 1%.

However, the average neck diameter values observed in plants grown in the preserved forest soil in the natural condition were higher (12.8%) than those of plants grown in the soil degraded by the tailings. In the sterile soil condition, there was no significant influence on neck diameter between the soil types, a fact that may possibly be related to the increase in manganese levels in the preserved forest soil after sterilization, making it close to the levels in the soil degraded by the tailings.

With regard to the interaction between soil condition and inoculation treatment (Table 7), the highest average neck diameter values were recorded in the natural condition when inoculated with *G. etunicatum* (4.93 mm).

Table 7 - Collar diameter (CD) of sabia plants (*Mimosa Caesalpiniaefolia* Benth.) subjected to two soil conditions (sterile and natural) and four inoculation treatments with AMF.

Soil Condition	Inoculation treatment			
	Control	*G. clarum*	*G. etunicatum*	Mix
			DC (mm)	
Sterile	1.44 bB	1.62 bB	4.43 aA	3.96 aA
Natural	4.07 aB	4.33 aB	4.93 aA	4.41 aB

[1]-Averages followed by equal letters in the row and in the columns do not differ by the Scott-Knott test at a significance level of 1%.

When analyzing each soil condition (sterile and natural), the plants grown in sterile soil and inoculated with *G. etunicatum* and Mix were significantly superior to those treated with *G. clarum* and the control. In relation to the natural condition, only the treatment with *G. etunicatum* had an influence, with an increase of 17.4% compared to the control. When comparing the AMF inoculation treatments between soil conditions, only the treatments with *G. etunicatum* and Mix showed no significant differences between the sterile and natural soil.

5.1.4. Root system length

With regard to root system length (RSC), it was found that the plants grown in the tailings-degraded soil were not influenced by inoculation with the AMF or by the barren and natural soil conditions (Table 8).

In plants grown in the preserved forest soil, the highest values for the length of the root system were observed in plants inoculated with *G. etunicatum* in the sterile soil condition, which promoted a 46% increase in this variable compared to the control. When comparing soil conditions, most of the plants grown in the natural condition were superior to the sterile condition, with the exception of plants inoculated with *G. etunicatum*. In general, the highest values for this variable were observed in the plants grown in the preserved forest soil under natural conditions.

Table 8 - Root system **length** (RSC) of sabia plants (*Mimosa Caesalpiniaefolia* Benth.) subjected to two types of soil (degraded by tailings and preserved forest), two conditions (sterile and natural) and four inoculation treatments with AMF.

Inoculation treatments (AMF)	Degraded by Tailings		Preserved Forest	
	Sterile	Natural	Sterile	Natural
		CSR (cm)		
Control	27.56 aA	31.06 aA	19.93 bB	31.12 aA
Glomus clarum	33.12 aA	30.81 aA	23.87 bB	36.43 aA
Glomus etunicatum	28.18 aA	31.31 aA	36.93 aA	29.68 aB
Mix	30.12 aA	28.75 aA	24.12 bB	32.75 aA

[1] - Averages followed by the same lowercase letter in the column for each soil condition and uppercase letters in the row for each soil type separately do not differ by the Scott-Knott test at 1% significance.

The high levels of Mn present in the soil degraded by the tailings may have inhibited the influence of the inoculation treatments on the length of the root system compared to the preserved forest. Similarly, Cipriani (2011) studying the behavior of sabià and *Acacia mangium* plants in soil under doses of arsenic also found no influence of AMF on the length of the root system. The same author also observed that as the dose of this metal increased, there was a greater reduction in the length of this variable.

On the other hand, Solis-Dominguez *et al.* (2011) working on the effect of arbuscular mycorrhizal fungi on algaroba plants grown in zinc mining soil, which also contains high levels of lead, observed significant differences in root growth when inoculated with AMF compared to the absence of inoculation. According to these authors, a commercial inoculum of AMF promoted greater growth in root length, even though there were no differences between the other two AMF studied (*G. intraradices* and a native inoculum). Similar results in the increase of root length were also observed by Chen *et al.* (2007) in plants of *P.* vittata, *T. repens* and *C. drummondii* colonized by

Glomus mosseae compared to those not colonized in copper mining soil.

5.1.5. relative chlorophyll index

In relation to the relative chlorophyll index (Table 9), it can be seen that plants grown in the soil degraded by the tailings only showed a significant difference between the inoculation treatments in the sterile condition, in which the highest averages were found for the Mix treatment (25.15), followed by *G. etunicatum* (24.03) and *G. clarum* (20.53), respectively, in relation to the control.

Table 9 - Relative chlorophyll index (CRI) of sabia plants (*Mimosa Caesalpiniaefolia* Benth.) subjected to two types of soil (degraded by tailings and preserved forest), two conditions (sterile and natural) and four inoculation treatments with AMF.

Inoculation treatments (AMF)	Degraded by Tailings		Preserved Forest	
	Sterile	Natural	Sterile	Natural
			IRC	
Control	14.81 cB	24.95 aA	17.57 bB	24.95 bA
Glomus clarum	20.53 bB	27.06 aA	16.00 bB	27.52 aA
Glomus etunicatum	24.03 aB	28.02 aA	26.00 aB	30.05 aA
Mix	25.15 aA	26.92 aA	25.17 aB	28.87 aA

[1] - Averages followed by the same lowercase letter in the column for each soil condition and uppercase letters in the row for each soil type separately do not differ by the Scott-Knott test at 1% significance.

When analyzing each inoculation treatment between the two soil conditions (barren and natural), it can be seen that, in general, the plants grown in the soil degraded by the tailings had the highest averages when grown under natural conditions, with the exception of the Mix inoculation treatment. With regard to the preserved forest soil, analyzing each soil condition in isolation, it can be seen that, in the sterile condition, both the Mix treatment (25.17) and *G. etunicatum* (26.00) were significantly superior to the treatment with *G. clarum* and the control. In the natural condition, the highest value for IRC was recorded when the plants were inoculated with *G. etunicatum*, followed by the Mix treatment and *Glomus clarum*, which promoted increases of 16.9%, 13.6% and 9.3% respectively, compared to the control.

The lower IRC observed in the control treatment and *G. clarum* in the sterile condition in both soils analyzed may be related to the appearance of toxicity symptoms in the two treatments mentioned, with these symptoms being more expressive in the plants of the control treatment grown in the soil degraded by the tailings under sterile conditions. According to Becerril *et al.* (2013), high concentrations of manganese in the leaves can contribute to a decrease in the photosynthetic rate and the appearance of brown necrotic spots on the leaves, which may have caused a reduction in the IRC.

Studies carried out by AL-AMRI (2013) with inoculation of AMF in *Vicia faba* plants grown in soils contaminated with metals from wastewater at different concentrations also showed similar results to the present work. This author observed higher IRC values in plants inoculated with AMF compared to the absence of inoculation. Results also found by Zhang *et al.* (2010) in maize plants grown in soils with different concentrations of lead and Latef *et al.* (2011) in pepper plants grown in soil contaminated with copper.

When comparing each inoculation treatment between the sterile and natural soil conditions, it can be seen that all the treatments in the natural condition were superior to the sterile condition, which could possibly be related to the nodulation of the sabia plants by native bacteria capable of leaving N_2 atmospheric. According to Figueiredo *et al.* (2011) the chlorophyll content measured with the SPAD chlorophyll meter correlates positively with the nitrogen concentration in the plant.

5.1.6. Aerial dry matter mass (MMSPA)

In plants grown in the soil degraded by the tailings, when analyzing each soil condition (sterile and natural), it was observed that the production of MMSPA only showed significant differences between the inoculation treatments in the sterile condition, which was more significantly influenced by the treatments with Mix and *G. etunicatum*, which promoted an increase in MMSPA of 96.6% and 95.5% compared to the absence of inoculation (Table 10). When analyzing the differences for each AMF inoculation treatment between the soil conditions (sterile and natural), plants grown in soil in the natural condition showed a significant increase in MMSPA production compared to the sterile condition, with the exception of the Mix inoculation treatment.

Table 10 - Mass **of** dry matter of the aerial part (MMSPA) of sabia plants (*Mimosa Caesalpiniaefolia* Benth.) submitted to two types of soil (degraded by tailings and preserved forest), two conditions (sterile and natural) and four inoculation treatments with AMF.

Inoculation treatments (AMF)	Degraded by Tailings		Preserved Forest	
	Sterile	Natural	Sterile	Natural
			MMSPA (g)	
Control	0.28 cB	7.20 aA	0.29 cB	8.82 bA
Glomus clarum	0.58 cB	8.48 aA	0.37 cB	9.90 bA
Glomus etunicatum	6.29 bB	8.26 aA	13.45 aA	12.21 aA
Mix	8.24 aA	7.43 aA	9.82 bB	12.23 aA

[1] - Means followed by the same lowercase letter in the column for each soil condition and uppercase letters in the row for each soil type separately do not differ by the Scott-Knott test at 1% significance.

Results in the increase of MMSPA production in soils with metals and inoculated with AMF were also observed in soybean plants with *Glomus macrocarpum* in soil with high manganese contents (CARDOSO *et al.*, 2003), 2003), algaroba plants inoculated with *Glomus intraradices* in soil containing lead (SOLÌS-DOMÌNGUEZ *et al.*, 2011) and *Brachiaria decumbens* Stapf. plants inoculated with *Scutellospora pellucida* in soil containing Zn, Cd, Cu and Pb (SILVA, 2006).

In relation to the plants grown in preserved forest soil, the highest MMSPA production was observed under the sterile condition, especially when inoculated with *G. etunicatum* (13.45 g) and Mix (9.82 g), while the plants inoculated with *G. clarum* showed similar results to the control. Under natural conditions, the highest values for this variable were significantly influenced by the Mix (12.3 g) and *G. etunicatum* (12.21 g) treatments. When comparing the MMSPA of each treatment of inoculation with AMF of the plants grown in the preserved forest soil between the soil conditions, it can generally be seen that the MMSPA under the natural condition was higher in all treatments compared to the sterile condition, with the exception of the *G. etunicatum* treatment.

The plants grown in the preserved forest soil showed higher values than those grown in the soil degraded by the tailings, a fact easily explained by the lower concentration of Mn present in this soil (Table 1), which resulted in greater plant growth and consequently greater biomass.

5.1.7. Mass of root dry matter (MMSR)

The absence of inoculation in sabia plants caused a decrease in MMSR production in both soils (tailings degraded and preserved forest) under both conditions (sterile and natural). In the tailings-degraded soil, under sterile conditions, the Mix treatment and *G. etunicatum* increased MMSR production by 96 % and 94.3 %, respectively, compared to the control treatment. While in the natural condition, *G. etunicatum*, *G. clarum* and Mix were the most efficient in promoting an increase in MMSR when compared to the absence of inoculation (Table 11). In general, when analyzing each AMF inoculation treatment between soil conditions, MMSR in the natural condition showed higher values than in the sterile condition, with the exception of the Mix treatment.

In the preserved forest soil, the treatments with *G. etunicatum* and Mix under sterile conditions were also the most efficient, promoting increases in MMSR of 97.5% and 95.6%, respectively, when compared to the control. Under natural conditions, the increases occurred with the inoculation *of G. etunicatum* (39.9%), Mix (37.6%) and *G. clarum* (33.7%). In general, the plants grown in the preserved forest soil showed higher MMSR values than the plants in the soil degraded by the tailings.

Table 11 - Root dry matter mass (RSM) of sabia plants (*Mimosa Caesalpiniaefolia* Benth.) subjected to two types of soil (degraded by tailings and preserved forest), two conditions (barren and natural) and four inoculation treatments with AMF.

Inoculation treatments (AMF)	Degraded by Tailings		Preserved Forest	
	Sterile	Natural	Sterile	Natural
		MMSR (g)		
Control	0.14 bB	2.98 bA	0.12 cB	3.02 bA
Glomus clarum	0.18 bB	4.54 aA	0.17 cB	4.56 aA
Glomus etunicatum	2.46 aB	4.31 aA	4.83 aA	5.03 aA
Mix	3.53 aA	3.76 aA	2.77 bB	4.84 aA

[1] - Averages followed by the same lowercase letter in the column for each soil condition and uppercase letters in the row for each soil type separately do not differ by the Scott-Knott test at 1% significance.

Similarly to the behavior of MMSPA, MMSR highlights the sensitivity of the mycorrhizal species and the production of MMSR to higher concentrations of metals, a fact confirmed by Siqueira *et al.* (1999) when they studied arbuscular mycorrhizae in the growth of tree seedlings in soil with an excess of heavy metals.

In general, in both soils, the treatments inoculated with AMF produced a greater amount of MMSR compared to the control. Jankong and Visoottiviseth (2008), when studying three plant species inoculated with AMF in soil contaminated with arsenic, observed that the MMSR of the species *Melastoma malabathricum* was significantly affected by 78.9% compared to the absence of inoculation. On the other hand, the same authors observed that the other two species studied were not affected by inoculation with AMF compared to the plants in the control treatment. It is clear that the benefits caused to mycorrhized plants depend greatly on the species of fungus and the plant involved in the association.

5.2. Spore density and mycorrhizal root colonization

Table 12 shows that only the interaction between the three factors was not significant for the spore density (SD) variable. In relation to root mycorrhizal colonization, all the isolated factors, as well as the interactions soil type x AMF inoculation treatment and soil condition x AMF inoculation treatment were significantly affected.

Table 12 - Summary of analysis of variance for spore density (SD) and mycorrhizal root colonization (MRC) for treatment factors and their interactions in *Mimosa Caesalpiniaefolia* Benth.

F. V.	G.L.	MEAN SQUARE	
		OF	CMR
TS (A)	1	$304{,}47^{**}$	$38{,}03^{**}$
CS (B)	1	$77{,}60^{**}$	$233{,}11^{**}$
AMF (C)	3	$23{,}28^{**}$	$79{,}09^{**}$
Int. (AxB)	1	$58{,}57^{**}$	$2{,}35^{ns}$
Int. (AxC)	3	$10{,}77^{**}$	$5{,}24^{**}$
Int. (BxC)	3	$4{,}86^{**}$	$64{,}53^{**}$
Int. (AxBxC)	3	$2{,}44^{ns}$	$0{,}87^{ns}$
Total	63	-	-
C.V.%	-	31,97	15,34

G.L - degree of freedom; (A) - soil type; (B) - soil condition; (C) - AMF inoculation treatment; CV - coefficient of variation; **, *, ns - significant by the Scott-Knott test at 1%, 5% and non-significant, respectively.

5.2.1. Spore density

For spore density (Table 13), the highest values were found in the preserved forest soil under the inoculation treatment with the AMF *Glomus etunicatum* (3893.12 spores/100 g of soil).$^{-1}$

Table 13 - Spore density (SD) of sabia plants (*Mimosa Caesalpiniaefolia* Benth.) subjected to two types of soil (degraded by tailings and preserved forest) and four inoculation treatments with AMF.

Soil type	Inoculation treatment			
	Control	*G. clarum*	*G. etunicatum*	Mix
	Spore density (sp. 100 g)$^{-1}$			
Degraded by Tailings	255.62 bA	412.50 bA	717.50 bA	499.37 bA
Preserved Forest	1515.93 aD	2255.62 BC	3893.12 aA	2901.87 aB

[1-]Averages followed by equal letters in the row and in the columns do not differ by the Scott-Knott test, at a significance level of 1%.

When analyzing each type of soil in isolation, it can be seen that there was only a significant influence between the inoculation treatments in the preserved forest soil, in which the AMF *G. etunicatum,* Mix and *G. clarum* promoted increases in spore density of 61 %, 47.7 % and 32.7%, respectively, compared to the control. In general, the density of spores in the soil degraded by the tailings showed a significant reduction compared to the preserved forest soil.

Schneider *et al.* (2012), studying the use of pteridophytes associated with arbuscular mycorrhizal fungi in soil contaminated with arsenic, found a reduction in the density of AMF spores in soil with tailings compared to less contaminated soil. Similar results were also observed by Klauberg-Filho *et al.* (2002), who found a higher density of spores in sites with less heavy metal contamination. According to Stürmer and Bellei (1994), this may be related to the better vegetative growth of the host and the higher level of mycorrhizal colonization in soils with lower levels of contamination.

Another hypothesis to be considered in the reduction of spore density in the soil degraded by the tailings is the fact that the plants are less developed due to the lower presence of organic matter and the high levels of Mn when compared to the preserved forest soil. These factors, according to Pawlowska and Charvat (2004), can limit spore germination, the development of extraradicular mycelium and sporulation of the fungus.

Taking into account the interaction between soil condition and AMF inoculation treatment (Table 13), it can be seen that in both the sterile and natural conditions there was a significant difference between the inoculation treatments, which was more significant in the natural condition with *G. etunicatum* (39.4%) compared to the absence of inoculation, while the Mix treatment and *G. clarum* were similar to the control. On the other hand, in the sterile condition all the inoculated treatments were significantly superior to the control. In general, the natural condition showed higher spore density values than the sterile soil.

Table 14 - Spore density (SD) of sabia plants (*Mimosa Caesalpiniaefolia* Benth.) subjected to two conditions (sterile and natural) and four inoculation treatments with AMF.

Soil Condition	Inoculation treatment			
	Control	*G. clarum*	*G. etunicatum*	Mix
		Spore density (sp. 100 g)$^{-1}$		
Sterile	0.00 bC	886.25 bB	1685.00 bA	1463.12 aA
Natural	1771.56 aB	1781.87 aB	2925.62 aA	1938.12 aB

"Means followed by equal uppercase letters in the row and lowercase letters in the columns do not differ by the Scott-Knott test, at a significance level of 1%.

It is important to point out that in the natural condition in the preserved forest soil, a much greater diversity of spores was found than in the soil degraded by the tailings. In the preserved forest soil, spores of the genus *Glomus* sp. and *Gigaspora* sp. were identified, while in the soil degraded by the tailings only the genus *Glomus sp.* was identified. Bezerra *et al.* (2011) reported that the genus *Glomus* has a high adaptive capacity to wide ranges of environmental conditions.

According to Souza *et al.* (2011), when it comes to the use of symbiotic microorganisms in phytoremediation programs in contaminated areas, the evaluation of AMF species that bring greater benefits when associated with a particular plant species can consequently increase its potential in the phytoremediation process.

5.2.2. Mycorrhizal root colonization

According to the soil type x AMF inoculation treatment interaction for mycorrhizal colonization (Table 15), it can be seen that in both soils (degraded by tailings and preserved forest) the treatments with *G. etunicatum* and Mix were the most significant in increasing this variable. Analyzing the soil degraded by the tailings, it is possible to see that the increases were in the order of 48.5% and 44.9%, respectively, for the Mix and *G. etunicatum* treatments in relation to the control. In the case of the preserved forest, the increases were 60.1%, 56.5% and 36.7%, respectively, for the *G. etunicatum*, Mix and *G. clarum* treatments.

Table 15 - Root mycorrhizal colonization (RMC) of sabia plants (*Mimosa Caesalpiniaefolia* Benth.) subjected to two types of soil (degraded by tailings and preserved forest) and four inoculation treatments with AMF.

Soil type	Inoculation treatment			
	Control	*G. clarum*	*G. etunicatum*	Mix
		Mycorrhizal Root Colonization (%)		
Degraded by Tailings	22.81 aB	27.56 bB	41.45 bA	44.30 bA
Preserved Forest	23.36 aC	36.94 aB	58.60 aA	53.75 aA

"Means followed by equal uppercase letters in the row and lowercase letters in the columns do not differ by the Scott-Knott test, at a significance level of 1%.

When analyzing the two types of soil in general, it can be seen that in the soil degraded by the tailings there was a reduction in the percentage of mycorrhizal root colonization of the plants in all inoculation treatments compared to the preserved forest soil.

However, despite this marked reduction in the tailings-degraded soil, the Mix and *G. etunicatum* treatments promoted an increase in mycorrhizal colonization compared to the control treatment, which may indicate a greater adaptability of both the plant species and the microorganisms to these conditions. Results similar to those found for this variable were also observed by Nogueira *et al.* (2003) when studying symbiotic efficiency and manganese toxicity in soybeans as a function of two types of soil. These authors observed that during the 12-week period, plants inoculated with the AMF *G. etunicatum and G. macrocarpum* showed greater colonization of the roots in soil with a higher concentration of Mn when compared to the absence of inoculation.

Becerril *et al.* (2013) when analyzing the impacts of manganese on the environment, interactions between soils, plants and arbuscular mycorrhizae, observed an absence of mycorrhizal colonization in *A. psilostachya* plants grown in mining waste soils, even though they had 44.7 spores/250 g of soil. In contrast, these authors reported that this same species grown in soil under native vegetation showed both colonization of the roots by AMF and a higher spore density compared to plants grown in Mn mining waste.

In relation to the interaction between soil condition and inoculation treatment (Table 16), the highest average CMR value was found in the natural condition with the *G. clarum* treatment (52.25 %).

Table 16 - Root mycorrhizal colonization (RMC) of sabia plants (*Mimosa Caesalpiniaefolia* Benth.) subjected to two conditions (sterile and natural) and four inoculation treatments with AMF.

Soil Condition	Inoculation treatment			
	Control	*G. clarum*	*G. etunicatum*	Mix
		Mycorrhizal Root Colonization (%)		
Sterile	0.00 bC	12.25 bB	48.81 aA	48.12 aA
Natural	46.18 aA	52.25 aA	51.25 aA	49.93 aA

[1]-Averages followed by equal letters in the row and in the columns do not differ by the Scott-Knott test, at a significance level of 1%.

Analyzing each soil condition (sterile and natural), it can be seen that there was only a significant difference between the inoculation treatments in the sterile condition, in which the *G. etunicatum* and Mix treatments were superior to *G. clarum* and the control. In general, plants grown under natural conditions showed a higher percentage of mycorrhizal root colonization when compared to sterile conditions. Results similar to these were also found by Aguiar *et al.* (2004) in algaroba plants and Carneiro *et al.* (2009) in alfalfa plants under natural conditions.

According to Smith and Read (1997), increased colonization is usually followed by stimulated plant growth, most often attributed to increased P absorption in soils with low P availability.

5.3. Mycorrhizal efficiency

In the summary of the analysis of variance in Table 17, it can be seen that all the isolated factors, as well as their interactions, had a significant effect on mycorrhizal efficiency.

Table 17- Summary of the analysis of variance for mycorrhizal efficiency (*MFE*) for the treatment factors and their interactions in *Mimosa Caesalpiniaefolia* Benth.

F. V.	G.L.	MEAN SQUARE
		EFM
TS (A)	1	8,88**
CS (B)	1	156,12**
AMF (C)	2	38,52**
Interaction (AxB)	1	8,24**
Interaction (AxC)	2	5,88**
Interaction (BxC)	2	38,20**
Interaction (AxBxC)	2	5,74**
Total	47	-
C.V.%	-	53,84

G.L - degree of freedom; (A) - soil type; (B) - soil condition; (C) - AMF inoculation treatment; CV - coefficient of variation; **, *, ns - significant by the Scott-Knott test at 1%, 5% and non-significant, respectively.

Analyzing the mycorrhizal efficiency of the three AMF inoculation treatments, it can be seen that both in the soil degraded by the tailings and in the preserved forest soil there was only a significant influence of the inoculation treatments in the barren condition (Table 18). In the tailings-degraded soil, the Mix treatment and *G. etunicatum* stood out from the control in this same condition, with maximum values of 2603.64 % and 1959.77 %, respectively.

Table 18 - Mycorrhizal efficiency (*MFE*) of sabia plants (*Mimosa Caesalpiniaefolia* Benth.) subjected to two types of soil (degraded by tailings and preserved forest), two conditions (sterile and natural) and three inoculation treatments with AMF.

Inoculation treatments (AMF)	Degraded by Tailings		Preserved Forest	
	Sterile	Natural	Sterile	Natural
		FSM (%)		
Glomus clarum	84.96 bA	28.57 aA	31.81 cA	23.35 aA
Glomus etunicatum	1959.77 aA	23.60 aB	4398.39 aA	46.53 aB
Mix	2603.64 aA	10.62 aB	3007.03 bA	44.88 aB

[1] - Averages followed by the same lowercase letter in the column for each soil condition and uppercase letters in the row for each soil type separately do not differ by the Scott-Knott test at 1% significance.

Similar results were also observed by Nogueira *et al.* (2003), working with the symbiotic efficiency of AMF in soybean plants in a soil with high manganese levels. The authors found that plants inoculated with the species *G. etunicatum* showed a higher percentage of mycorrhizal efficiency compared to plants in the control treatment over a 12-week period. Andrade *et al.* (2003) also reported an increase in the mycorrhizal efficiency of *G. macrocarpum* in the production of dry matter in soybean plants grown in soil containing lead. According to the authors, inoculation with this fungus promoted a 2.3-fold increase compared to non-mycorrhized plants.

In general, the highest percentage of mycorrhizal efficiency in both soils was observed in the sterile condition, with the values for this variable in the preserved forest soil being higher than those in the soil degraded by the tailings. The greater expressiveness of the AMF in the barren soil is possibly related to the mycorrhizal dependence of the species studied, as well as the elimination of the soil's native microbiota, which consequently may have reduced the competitiveness of the introduced AMF in relation to the native ones.

5.4. N, P and K content in the aerial part

According to the summary of the analysis of variance (Table 19), only the nutritional levels of N and P were significantly influenced by the single factors soil condition and AMF inoculation treatment. For the single factor soil type, only phosphorus did not show a significant influence. For the soil type x soil condition and soil type x AMF inoculation treatment interactions, there was no significant influence on any of the nutritional variables analyzed. With regard to the soil condition x AMF inoculation treatment interaction, only the phosphorus (P) content was significantly influenced. In the interaction between the three factors studied, only potassium did not show a significant influence.

Table 19- Summary of the analysis of variance for the nitrogen (N), phosphorus (P) and potassium (K) contents in the aerial part for the treatment factors and their interactions in *Mimosa Caesalpiniaefolia* Benth.

F. V.	G.L.	MEAN SQUARE		
		N	P	K
TS (A)	1	22,04**	0,27^{ns}	13,53**
CS (B)	1	47,91**	31,31**	0,15^{ns}
AMF (C)	3	22,16**	20,80**	1,61^{ns}
Interaction (AxB)	1	0,37^{ns}	3,22^{ns}	0,59^{ns}
Interaction (AxC)	3	1,19^{ns}	2,58^{ns}	0,26^{ns}
Interaction (BxC)	3	2,08^{ns}	20,07**	2,24^{ns}
Interaction (AxBxC)	3	3,08*	5,48**	0,72^{ns}
Total	63	-	-	-
C.V.%	-	10,26	13,13	19,33

G.L - degree of freedom; (A) - soil type; (B) - soil condition; (C) - AMF inoculation treatment; CV - coefficient of variation; **, *, ns - significant by the Scott-Knott test at 1%, 5% and non-significant, respectively.

5.4.1. Nitrogen

For the nitrogen content in the aerial part (Table 20), when analyzing each soil condition (barren and natural), it was found that the plants grown in the soil degraded by the tailings showed a significant difference only in the inoculation treatments in the natural condition, in which the Mix treatment promoted an increase of 24.9% compared to the absence of inoculation. Meanwhile, plants grown in preserved forest soil showed a significant effect of AMF in both soil conditions (barren and natural).

In the sterile condition, the highest N levels were found in the presence of the Mix treatment (14.61 g kg^{-1}) followed by *G. etunicatum* (12.63 g kg^{-1}), while the treatment with *G. clarum* showed similar results to the control. Under natural conditions, the highest N levels in the aerial part of the plants were also caused by the Mix treatment, which was responsible for an increase of 18.6% compared to the control treatment. In general, it can also be seen that in both soils (tailings degraded and preserved forest), plants grown under natural conditions had higher N levels compared to the barren condition.

Table 20- Nitrogen (N) content in sabia plants (*Mimosa Caesalpiniaefolia* Benth.) subjected to two types of soil (degraded by tailings and preserved forest), two conditions (sterile and natural) and four inoculation treatments with AMF.

Inoculation treatments (AMF)	Degraded by Tailings		Preserved Forest	
	Sterile	Natural	Sterile	Natural
		Nitrogen (g kg)$^{-1}$		
Control	10.13 aA	11.53 bA	9.84 cB	12.88 bA
Glomus clarum	9.35 aB	11.55 bA	10.22 cB	13.43 bA
Glomus etunicatum	10.25 aA	11.62 bA	12.63 bA	12.99 bA
Mix	10.99 aB	15.37 aA	14.61 aA	15.83 aA

[1] - Averages followed by the same lowercase letter in the column for each soil condition and uppercase letters in the row for each soil type separately do not differ by the Scott-Knott test at 5% significance.

According to Shukla *et al.* (2012) nitrogen is considered one of the most important essential elements for plant growth and development, especially in areas where this nutrient is limited. In these areas, the presence of arbuscular mycorrhizal fungi can favor the absorption of various elements considered essential, such as N (JOHNSON, 2010). According to Hodge *et al.* (2001) the extraradicular hyphae of AMF are capable of absorbing ammonium, nitrate and some amino acids from their surroundings and translocating the N to the plant. Another benefit of AMF is related to a more adequate supply of P, which can benefit the process of biological nitrogen fixation, which is highly demanding of energy in the form of ATP (JESUS *et al.*, 2005).

Some studies have demonstrated the efficiency of AMF in accumulating nitrogen in plants under stressful conditions, as in the case of areas contaminated with metals. Aram *et al.* (2013), when studying the effect of arbuscular mycorrhizal fungi on nitrogen concentration in *Berseem Clover* in soil contaminated with cadmium, reported a greater increase in N accumulation in plants colonized by AMF compared to non-inoculated plants. These results also corroborate those obtained by Guo *et al.* (2013)

when working with inoculation of maize and sorghum plants grown on mine tailings containing rare earth metals. According to these authors, inoculation with the AMF *G. versiforme* significantly increased the N content in the aerial part of both plants compared to the plants in the control treatment (uninoculated).

5.4.2. Phosphorus

In terms of the phosphorus (P) content of the aerial part (Table 21), it was observed that, in the soil degraded by the tailings and preserved forest, there was only a significant influence between the inoculation treatments in the sterile condition. The absence of a significant effect in the inoculation treatments under natural conditions may be related to the greater competition from the native AMF compared to the inoculated ones, as demonstrated by Costa *et al.* (2005).

The plants grown in the soil degraded by the tailings in the sterile condition showed a greater accumulation of P when inoculated with *G. etunicatum* (0.84 g kg^{-1}) followed by Mix (0.67 g kg^{-1}), while the treatment with *G. clarum* showed similar results to the control. The increases in P content promoted by *G. etunicatum* and Mix were 34.1% and 14.9% compared to the absence of inoculation.

Table 21- Phosphorus (P) content in sabia plants (*Mimosa Caesalpiniaefolia* Benth.) subjected to two types of soil (degraded by tailings and preserved forest), two conditions (sterile and natural) and four inoculation treatments with AMF.

Inoculation treatments (AMF)	Degraded by Tailings		Preserved Forest	
	Sterile	Natural	Sterile	Natural
	Phosphorus (g kg $)^{-1}$			
Control	0.57 cA	0.57 aA	0.47 bA	0.58 aA
Glomus clarum	0.54 cA	0.57 aA	0.51 bA	0.52 aA
Glomus etunicatum	0.84 aA	0.53 aB	0.90 aA	0.55 aB
Mix	0.67 bA	0.64 aA	0.93 aA	0.56 aB

[1] - Averages followed by the same lowercase letter in the column for each soil condition and uppercase letters in the row for each soil type separately do not differ by the Scott-Knott test at 1% significance.

On the other hand, in the preserved forest soil under sterile conditions, the Mix treatment and *G. etunicatum* showed the same effectiveness in P accumulation. These same treatments caused increases of 49.4% and 47.7%, respectively, compared to the absence of inoculation. In general, in both soils, the highest levels of P accumulation were observed in the barren condition, and these values were higher in plants grown in the preserved forest soil, which was to be expected due to the better nutritional composition of the soil. In general, the higher accumulations of P in the aerial part of the plants inoculated with *G. etunicatum* and Mix may explain their better development compared to the plants in the control treatment. According to Hipler and Moreira (2013), the greater accumulation of P in the aerial part of plants associated with AMF can be attributed to an increase in the absorption surface promoted by the hyphae of these microorganisms.

Nogueira *et al.* (2007), studying the influence of mycorrhizal fungi and bacteria on the extraction and absorption of iron and manganese in soybean plants, reported a significant increase in the phosphorus content in the aerial part of plants colonized by *Glomus etunicatum* and *Glomus macrocarpum* compared to non-inoculated plants. Similar results were also observed by Chen *et al.* (2007) in four plant species inoculated with AMF and grown on copper mining tailings, and by Liu *et al.* (2015) in *Solanum nigrum* plants inoculated with *G. versiforme* in soil contaminated with cadmium.

5.4.3. Potassium

With regard to potassium (K) content, Table 22 shows that the plants grown in the preserved forest soil had the highest potassium accumulation (12.51 g kg^{-1}) in the aerial part, an increase of 16.3% compared to the plants grown in the soil degraded by the tailings.

Table 22- Potassium (K) content in sabia plants (*Mimosa Caesalpiniaefolia* Benth) subjected to two types of soil (degraded by tailings and preserved forest).

Soil type	Potassium (g kg^{-1})
Degraded by Tailings	10,47 b
Preserved Forest	12,51 a

[1] - Averages followed by the same lowercase letter in the column between soil types do not differ by the Scott-Knott test at 1% significance.

It is possible that the greater accumulation of potassium in the aerial part of the plants grown in the preserved forest soil may be related to a better nutritional condition, as well as the fact that it has a higher pH value (Table 1), which may consequently increase the availability of K for the plants. Similar results were also observed by Melo *et al.* (2015) in maize plants grown under different fertility management.

5.5. Mn content in the aerial part and root

According to the summary of the analysis of variance (Table 23), all the factors and their interactions had a significant effect on the manganese (Mn) content in the aerial part and root.

Table 23- Summary of the analysis of variance for manganese (Mn) content in the aerial part and root for the treatment factors and their interactions in *Mimosa Caesalpiniaefolia* Benth.

F.V.	G.L.	MEAN SQUARE	
		Mn air part	Mn root
TS (A)	1	1827,46**	106,26**
CS (B)	1	3254,85**	44,63**
AMF (C)	3	18,15**	4,28**
Int. (AxB)	1	928,87**	18,07**
Int. (AxC)	3	3,13*	6,54**
Int. (BxC)	3	6,66**	9,22**
Int. (AxBxC)	3	6,46**	3,29*
Total	63	-	-
C.V.%	-	9,71	24,06

G.L - degree of freedom; (A) - soil type; (B) - soil condition; (C) - inoculation treatment with AMF; CV - coefficient of variation; **, *, ns - significant by the Scott-Knott test at 1%, 5% and non-significant levels, respectively.

5.5.1. Manganese in the aerial part and root

As for the manganese (Mn) content in the aerial part (Table 24), it can be seen that, in both the soil degraded by the tailings and the preserved forest, there was only a significant effect of the inoculation treatments in the sterile condition. Plants inoculated in both soils (degraded by tailings and preserved forest) in the sterile condition inoculated with *G. etunicatum* provided the lowest values in the accumulation of Mn in the aerial part when compared to plants in the control treatment. Despite the significant absence of AMF in the natural condition, the treatment with *G. etunicatum* also resulted in lower values compared to the absence of inoculation. Analysing the differences between the conditions (sterile and natural), it can be seen that, both in the soil degraded by the tailings and in the preserved forest, the sterilization of the soil significantly influenced a greater accumulation of Mn in the plants grown under sterile conditions.

Table 24 - Manganese (Mn) content in the aerial part of sabia plants (*Mimosa Caesalpiniaefolia* Benth.) subjected to two types of soil (degraded by tailings and preserved forest), two conditions (barren and natural) and four inoculation treatments with AMF.

Inoculation treatments (AMF)	Degraded by Tailings		Preserved Forest	
	Sterile	Natural	Sterile	Natural
		Mn in the aerial part (mg kg)[-1]		
Control	1064.86 aA	196.28 aB	364.95 aA	82.94 aB
Glomus clarum	1002.50 bA	173.80 aB	348.65 aA	64.27 aB
Glomus etunicatum	998.41 bA	149.39 aB	169.66 bA	51.74 aB
Mix	1062.71 aA	210.45 aB	402.00 aA	54.50 aB

[1] - Averages followed by the same lowercase letter in the column for each soil condition and uppercase letters in the row for each soil type separately do not differ by the Scott-Knott test at 1% significance.

Results in the literature that report a reduction in metallic elements in the area of plants inoculated with arbuscular mycorrhizal fungi are common. In the case of manganese, some authors attribute this reduction to a greater accumulation of P in plants colonized by AMF, which can complex with Mn, reducing its activity in the plant (FOY *et al.*, 1978; NOGUEIRA; CARDOSO, 2002). According to Chen *et al.* (2007), in contaminated soils such as mining areas there is a limited supply of nutrients considered essential to plants in combination with an excess of metals, with mycorrhizal inoculation the host plant can obtain greater absorption of nutrients and consequently improve its resistance to metal contamination. However, the effects provided by AMF on metal absorption and consequent accumulation in their hosts can vary greatly, depending on the fungal inoculum, host plant and environment (MAN *et al.*, 2013).

Nogueira *et al.* (2004), working with manganese toxicity and mycorrhizal fungi in soybean plants, observed that, at 90 days, soybean plants inoculated with *G. macrocarpum* and plants inoculated with *G. etunicatum* showed, respectively, a greater reduction in the accumulation of manganese in the aerial part compared to plants in the control treatment. Also according to these authors, the fact that plants inoculated with *G. macrocarpum* and *G. etunicatum* differed in Mn concentration may be related to differences intrinsic to each species of AMF. Similar results were also observed by Nogueira; Cardoso (2002) when they studied microbial interactions in the availability and absorption of manganese by soybeans.

The roots behaved in the opposite way to the aerial part (Table 25), in general with lower Mn content in the roots of non-mycorrhized plants and higher in mycorrhized plants, with the exception of plants grown in the preserved forest soil under natural conditions. However, it should be noted that only plants grown in soil degraded by tailings under sterile conditions showed a significant difference between the inoculation treatments, with these results being more expressive in the presence of Mix, followed by *G. etunicatum*, which provided increases of 67.1 % and 52.6 %, respectively, compared to the control. In general, it can be seen that the accumulation of Mn in the roots was higher compared to the accumulation in the aerial part of the plants.

When considering the differences between soil conditions, it was observed that plants grown in soil degraded by tailings under natural conditions showed greater accumulation of Mn in the roots when compared to the barren condition, with the exception of Mix. As for the preserved forest, only the control in the natural condition was superior to the barren condition.

Table 25 - Manganese (Mn) content in the roots of sabia plants (*Mimosa Caesalpiniaefolia* Benth.) subjected to two types of soil (degraded by tailings and preserved forest), two conditions (sterile and natural) and four inoculation treatments with AMF.

Inoculation treatments (AMF)	Degraded by Tailings		Preserved Forest	
	Sterile	Natural	Sterile	Natural
	Mn in the root $(mg\ kg)^{-1}$			
Control	4609.6 cB	13492.0 aA	3610.4 aB	7837.2 aA
Glomus clarum	4660.9 cB	13997.9 aA	6259.7 aA	6221.9 aA
Glomus etunicatum	9743.1 bB	13674.1 aA	4805.5 aA	6928.6 aA
Mix	14036.9 aA	14040.8 aA	6102.2 aA	4712.3 aA

[1] - Averages followed by the same lowercase letter in the column for each soil condition and uppercase letters in the row for each soil type separately do not differ by the Scott-Knott test at 5% significance.

The greater amount of manganese accumulated in the roots under both soil conditions (sterile and natural) may explain the lower concentration of Mn accumulated in the aerial part, especially under natural conditions. In general, the plants accumulated more Mn in the roots than in the aerial part, which indicates the possibility of using this plant in phytostabilization programs. According to Santos *et al.* (2007) and Santibânez *et al.* (2008) phytostabilizing plants are characterized by absorbing a large amount of a metal, keeping it mainly in the roots. Species with these properties can facilitate the process of immobilizing the metal, thus reducing its availability to other organisms and the leaching of the contaminant into uncontaminated areas. It should also be noted that some metals are able to easily establish bonds with organic compounds in the roots, forming more stable complexes that are less toxic to the environment (ACCIOLY; SIQUEIRA, 2000; MELO *et al.*, 2009; VENDRUSCOLO, 2013).

Cardoso *et al.* (2003), studying the absorption and translocation of manganese in mycorrhized soybean plants, under increasing doses of this nutrient, reported the protective effect of the *Glomus macrocarpum* species in protecting plants against excess manganese. According to the authors, the plants associated with this species of AMF increased the Mn content in the roots and therefore its translocation was reduced to the aerial part. Similar results were also observed by Souza *et al.* (2011) in *Stizolobium aterrimum* plants inoculated with *G. etunicatum* in soil contaminated with lead.

5.6. Mn translocation factor

The Mn translocation factor (FT) ranged from 0.011 to 0.236 for the tailings-degraded soil and from 0.08 to 0.103 for the preserved forest (Table 25). The results found in this study generally show that plants inoculated with *G. etunicatum* had a lower capacity to translocate Mn to the aerial part in both soils (tailings degraded and preserved forest) and conditions (barren and natural) compared to the control treatments. It can also be seen that the translocation factor in plants grown under natural conditions was lower when compared to plants grown under barren conditions, which was to be expected given the higher concentration of this element in the roots under these conditions (natural). Above all, the values found in this study show a restriction in the translocation of Mn to the aerial part, since only values greater than 1 are considered high (DENG *et al.*, 2004).

Table 26 - Translocation factor (FT) in sabia plants (*Mimosa Caesalpiniaefolia* Benth.) submitted to two types of soil (degraded by tailings and preserved forest), two conditions (sterile and natural) and four inoculation treatments with AMF.

Inoculation treatments (AMF)	Degraded by Tailings		Preserved Forest	
	Sterile	Natural	Sterile	Natural
	Translocation Factor (FT)			
Control	0,236 ± 0,015	0,014 ± 0,002	0,103 ± 0,009	0,012 ± 0,003
Glomus clarum	0,217 ± 0,010	0,012 ± 0,002	0,057 ± 0,007	0,016 ± 0,005
Glomus etunicatum	0,103 ± 0,002	0,011 ± 0,002	0,035 ± 0,001	0,008 ± 0,002
Mix	0,076 ± 0,002	0,015 ± 0,002	0,077 ± 0,017	0,012 ± 0,002

[1] - Data presented as mean ± standard error (n=4).

With regard to the translocation factor, plants can be classified as tolerant when FT<1 and Mn accumulators when FT>1 (SANTILLAN *et al.*, 2010). Thus, the results obtained in this study demonstrate the tolerance capacity of sabia plants, especially when inoculated with *G. etunicatum*, once again reinforcing the possibility of using them in phytostabilization programs. Also according to Santillan *et al.* (2010), the tolerance capacity of plants with this characteristic may be related to some exclusion strategy of the plant itself, which allows it to form more stable Mn complexes in its roots, resulting in limited translocation to the aerial part. Andrade *et al.* (2009) reported that the reduction in the availability of metals by AMF is related to their greater immobilization in intra- and extraradicular symbiotic structures, which can consequently reduce the translocation of these elements to the aerial part.

Carneiro *et al.* (2001), working with herbaceous species, observed that the concentration of cadmium in the roots was higher in plants inoculated with AMF and about three times higher in relation to the aerial part, thus showing the lower translocation of these metals in inoculated plants.

6. CONCLUSIONS

Cultivation in soil degraded by manganese mining tailings reduced the development of sabià plants when compared to those cultivated in preserved forest soil, regardless of the soil condition and inoculation treatments.

Inoculation with *G. etunicatum* and Mix favored the development of plants in sterile conditions, regardless of the type of soil, increasing their ability to tolerate Mn.

The high levels of Mn in the roots indicate that sabia (*Mimosa Caesalpiniaefolia* Benth.) can be effective in the process of phytostabilization in areas degraded by Mn mining.

REFERENCES

ABOU, M.; SYMEONIDIS, L.; HATZISTAVROU, E.; YUPSANIS, T. Nucieoiytic activities and appearance of a new DNase in relation to nickel and manganese accumulation in *Alyssum mûrale*. **Journal of Plant Physiology**, v.159, p.1087-1095, 2002.

ACCIOLY, A. M. A.; SIQUEIRA, J. O. Chemical contamination and soil bioremediation. in: NoVAis, R. F.; ALVAREZ V.; V. H.; sCHAEFER, C. E. G. R., eds.

Topics in soil science. Viçosa, Brazilian Society of Soil Science, 2000. p. 299-352.

AGuiAR, R. L. F. de; MAiA, L. C.; sALCEDo, i. H.; sAMPAio, E. V. de s. B. Interaction between arbuscular mycorrhizal fungi and phosphorus in the development of algaroba [*Prosopis juliflora* (Sw) DC]. **Revista Arvore**, v. 28, n. 4, p. 589-598, 2004.

AL-AMRi, s. M. The functional roles of arbuscular mycorrhizal fungi in improving growth and tolerance of *Vicia faba* plants grown in waste water contaminated soil.

African Journal of Microbiology Research, v. 7, n.35, p. 4435-4442, 2013.

ALENCAR, F. H. H. de. **Forage potential of the sabià species (*Mimosa caesalpiniifolia* Benth.) and its resistance to subterranean termites**. 2006. 61 f. Dissertation (Master's Degree in Animal Science) - Federal University of Campina Grande, Patos.

ANDRADE, S. A. L.; ABREu, C. A.; ABREu, M. F.; SiLVEiRA, A. P. D. interaction of lead, soil base saturation and arbuscular mycorrhiza on the growth and mineral nutrition of soybeans. **Revista Brasileira de Ciência do Solo**, v. 27, n. 5, p. 945-954, 2003.

ANDRADE, S. A. L.; GRATAo, P. L.; SCHiAViNATo, M. A.; SiLVEiRA, A. P. D.; AZEVEDo, R. A.; MAZZAFERA, P. Zn uptake, physiological response and stress attenuation in mycorrhizal jack bean growing in soil with increasing Zn concentrations. **Chemosphere**, v. 75, n.10, p.1363-1370, 2009.

ARAM, H.; GoLCHiN, A. The effects of arbuscular mycorrhizal fungi on nitrogen concentration of berseem clover in contaminated soil with cadmium. **Journal of Chemical Health Risks**, v. 3, n. 2, p. 35-38, 2013.

ARAUJO, I. C. S.; COSTA, M. C. G. Biomass and nutrient accumulation pattern of leguminous tree seedlings grown on mine tailings amended with organic waste.

Ecological Engineering, v.60, p.254-260, 2013.

BECERRIL, F. R.; VAZQUEZ, L. V. J.; CERVANTES, S. C. H.; SANDOVAL, O. A. A.; CORREA, G. V.; CHAVEZ, E. C.; ESPINDOLA, I. P. M.; HERRERA, A. E.; GONZALEZ, F. L. Impacts of manganese mining activity on the environment: interactions among soil, plants, and arbuscular mycorrhiza. **Archives of Environmental Contamination and Toxicology**, v.64, p.219-227, 2013.

BEZERRA, M. E. de J.; LACERDA, C. F. de; SOUZA, G. G. de; GOMES, V. F. F.; MENDES FILHO, P. F. Biomass, microbial activity and AMF in maize/corn-bean crop rotation using saline waters. **Revista Ciência Agronòmica**, v. 41, n. 4, p. 562-570, 2010.

BRANCHES, A. M. B.; RODRIGUES, V. M. Analysis of phytoremediation as a method for recovering areas degraded by mining. In: CONGRESSO BRASILEIRO DE MINA A CÉU ABERTO, 6., 2010, Belo Horizonte. **Proceedings...** Belo Horizonte: VI Brazilian Congress on Open-cast Mining, 2010.

BREMNER, J. M.; MULVANEY, C. S. Total nitrogen. *In* : PAGE , A. L. (Ed.). **Methods of soil analysis**. Madison: American Society of Agronomy, 1982. p. 595624.

BUCKMAN, H. O.; BRADY, N. C., 1974. Nature and properties of soils. Livraria Freitas Bastos. Rio de Janeiro, 594 p.

BURITY, H. A.; LYRA, M. do C. C. P. de; SOUZA, E. S. de; MERGULHÂO, A. C. do E. S.; SILVA, M. L. R. B. da. Effectiveness of inoculation with rhizobium and arbuscular mycorrhizal fungi on sabia seedlings subjected to different levels of phosphorus. **Pesquisa Agropecuària Brasileira**, v. 35, n. 4, p. 801-807, 2000.

CAIRES, S. M. de; FONTES, M. P. F.; FERNANDES, R. B. A.; NEVES, J. C. L.;

FONTES, R. L. F. Development of pink cedar seedlings in copper-contaminated soil: tolerance and potential for soil phytostabilization. **Revista Arvore**, v.35, n.6, p.1181-1188, 2011.

CARDOSO, E. J. B. N.; NAVARRO, R. B.; NOGUEIRA, M. A. Absorption and translocation of manganese by mycorrhized soybean plants, under increasing doses of this nutrient. **Revista Brasileira de Ciência do Solo**, v. 27, n. 3, p.415-423, 2003.

CARNEIRO, M. A. C.; SIQUEIRA, J. O.; MOREIRA, F. M. de S. Establishment of herbaceous plants in soil contaminated with heavy metals and inoculation of arbuscular mycorrhizal fungi. **Pesquisa Agropecuària Brasileira**, v. 36, n. 12, p. 1443-1452, 2001.

CARNEIRO, R. F. V.; EVANGELISTA, A. R.; ARAÙJO, A. S. F. Vegetative growth and nutrient acquisition by alfalfa in response to mycorrhiza and phosphorus doses. **Revista Brasileira de Ciências Agrârias**, v. 4, n. 3, p. 267-273, 2009.

CARVALHO, P. E. R. Circular Tècnica 135, **Sabià *Mimosa Caesalpiniaefolia*.**

EMBRAPA FORESTS. Colombo, PR November, 2007.

CASTRO, C. G. **Study of the use of manganese processing waste by the ceramics industry**. 2011. 107 f. Dissertation (Master's in Materials Engineering) - Federal University of Ouro Preto, Ouro Preto.

CHAVES, L. H. G.; MESQUITA, E. F.; ARAUJO, D. L.; FRANÇA, C. P. Accumulation and distribution of copper and zinc in castor bean cultivar BRS paraguaçu and plant growth. **Engenharia Ambiental: pesquisa e tecnologia**, v.7, n.3, p.263-277, 2010.

CHEN, B. D.; ZHU, Y. G.; DUAN, J.; XIAO, X. Y.; SMITH, S. E. Effects of the arbuscular mycorrhizal fungus *Glomus mosseae* on growth and metal uptake by four plant species in copper mine tailings. **Environmental Pollution**, v. 147, n. 2, p. 374380, 2007.

CHEN, X.; WU, C.; TANG, J.; HU, S. Arbuscular mycorrhizae enhance metal lead uptake and growth of host plants under a sand culture experiment. **Chemosphere**, v.60, p.665-671, 2005.

CHRISTIE, P.; LI, X.; CHEN, B. Arbuscular mycorrhiza can depressa translocation of zinc to shoots of host plants in soils moderately polluted with zinc. **Plant and Soil**, v.261, p.209-217, 2004.

CIPRIANI, H. N. **Morphophysiological responses of *Acacia mangium* Willd. and *Mimosa caesalpiniaefolia* Benth. plants, inoculated with rhizobium and arbuscular mycorrhiza, under the effect of arsenic**. 2011. 84 f. Dissertation (Master's Degree in Soils and Plant Nutrition) - Federal University of Viçosa, Viçosa.

CLEOLIN, D. **Effects of chronic manganese exposure on adult male mice**. 2010. 57 f. Dissertation (Master's Degree in Cellular and Structural Biology) - Federal University of Viçosa, Viçosa.

COSTA, C. M. C.; CAVALCANTE, U. M. T.; GOTO, B. T.; SANTOS, V. F. dos; MAIA, L. C. Arbuscular mycorrhizal fungi and phosphate fertilization in mangabeira seedlings. **Pesquisa Agropecuâria Brasileira**, v. 40, n. 3, p.225-232, 2005.

COSTA, G. S.; FRANCO, A. A.; DAMASCENO, R. N.; FARIA, S. M. Nutrient uptake by litter in a degraded area revegetated with tree legumes. **Revista Brasileira de Ciência do Solo**, v. 28, n.5, p.919-927, 2004.

D'AGOSTINO, L. F. **Mining tailings dam beaches: characteristics and sedimentation analysis**. 2008. 374 f. Thesis (Doctorate in Mineral Engineering) - Polytechnic School of the University of Sao Paulo, Sao Paulo.

DECHEN, A. R.; NACHTIGALL, G.R. Micronutrients. In: FERNANDES, M.S. (Ed.). **Mineral nutrition of plants**. Viçosa: Brazilian Society of Soil Science, 2006. p.327-354.

DEL VAL, C.; BAREA, J. M.; AGUILAR, C. A. Assessing the tolerance to heavy metals of arbuscular mycorrhizal fungi isolated from sewage sludge-contaminated soils. **Applied Soil Ecology**, v.11, p. 261-269, 1999.

DENG, H.; YE, Z. H. WONG, M. H. Accumulation of lead, zinc, copper and cadmium by 12 wetland plant species thriving in metal-contaminated sites in China.

Environmental Pollution, v. 132, n. 1, p. 29-40, 2004.

National Department of Mineral Production - DNPM. Mineral summary. Brasilia, 2013. Available at:

<https://sistemas.dnpm.gov.br/publicacao/mostra_imagem.asp?IDBancoArquivoArquivo=8992>. Accessed on: September 10, 2014.

EMBRAPA. **Manual of soil analysis methods**. Rio de Janeiro. 2. ed. rev.

Current. EMBRAPA, 1997. 212p.

EPSTEIN, E., 1975. **Plant mineral nutrition - principles and perspectives.**

Editora Universidade de Sao Paulo. 341 p.

FIGUEIREDO, L. A.; SILVA, D. H. da; BOARETTO, A. E.; RIBEIRINHO, V. S. Effect of glyphosate on^{15} N fixation in transgenic and conventional soybeans.

Revista Brasileira de Oleaginosas e Fibrosas, v.15, n. 1, p. 27-36, 2011.

FIGUEIRÔA, J. M. de; PAREYN, F. G. C.; DRUMOND, M.; ARAÙJO, E. de L.

(Eds.) **Species of northeastern flora of potential economic importance**. Recife: APNE, 2005. p.101-133.

FOY, C. D.; CHANEY, R. L.; WHITE, M. C. The physiology of metal toxicity in plants. **Annual Review of Plant Physiology**, v.29, p.511-566, 1978.

FRANCO, A. A.; CAMPELLO, E. F. C.; CAMPELLO, E. F. C.; SILVA, E. M. R.; FARIA, S. M. **Revegetation of Degraded Soils**. Technical communication. N° 09, Oct./92, p. 1-9 Dec./92 Ver. Mad. Embrapa Agrobiologia, Dec. 1992. 9p.

GERDEMANN, J. W.; NICHOLSON, T. H. Spores of

mycorrhizae *Endogone* species extracted from soil by wet sieving and decanting. **Transactions of the British Mycological Society**, v. 46, p. 235-244, 1963.

GONÇALVES, E.; SERFATY, A. **Perfil analitico do manganês**. Brasilia: National Department of Mineral Production/DNPM, 1976. Bulletin no. 37. 149p.

GRACE, C.; STRIBLEY, D. P. A safer procedure for routine staining of vesicular-arbuscular mycorrhizal fungi. **Mycological Research**, v. 95, n. 10, p. 1160-1162, 1991.

GUO, W.; ZHAO, R.; ZHAO, W.; FU, R.; GUO, J.; BI, N.; ZHANG, J. Effects of arbuscular mycorrhizal fungi on maize (*Zea mays* L.) and sorghum (*Sorghum bicolor* L. Moench) grown in rare earth elements of mine tailings. **Applied Soil Ecology**, v. 72, p. 85-92, 2013.

HIPPLER, F. W. R.; MOREIRA, M. Mycorrhizal dependence of peanut under phosphorus doses. **Soils and Plant Nutrition**, v. 72, n. 2, p. 184-191, 2013.

HODGE, A.; CAMPBELL, C. D.; FITTER, A. H. Fitter. An arbuscular mycorrhizal fungus accelerates decomposition and acquires nitrogen directly from organic material. **Nature**, v. 413, n. 6853, p. 297-299, 2001.

BRAZILIAN MINING Institute. **Exploration requests grow 374% in Ceará**. 2010. Available at: <http://www.ibram.org.br/>. Accessed on: August 20, 2014.

JACOBI, A. V. **Physiological changes caused by different concentrations of manganese in soybean cultivation: a bibliographical survey**. 2014. Monograph (Specialization in Soil Fertility and Plant Mineral Nutrition) - University of Cuiabà, Cuiabâ.

JANKONG, P.; VISOOTTIVISETH, P. Effects of arbuscular mycorrhizal inoculation on plants growing on arsenic contaminated soil. **Chemosphere**, v. 72, n. 7, p. 1092-1097, 2008.

JESUS, E. da C.; SCHIAVO, J. A.; FARIA, S. M de. Dependence on mycorrhizae for the nodulation of tropical tree legumes. **Revista Arvore**, v. 29, n. 4, p. 545552, 2005.

JOHNSON, N. C. Resource stoichiometry elucidates the structure and function of arbuscular mycorrhizas across scales. **New Phytologist**, v. 185, n. 3, p. 631-647, 2010.

KLAUBERG-FILHO, O.; SIQUEIRA, J. O.; MOREIRA, F. M. S. Arbuscular mycorrhizal fungi in soils polluted with heavy metals. **Revista Brasileira de Ciência do Solo**, v. 26, n. 1, p. 125-134, 2002.

KOSKE, R. E.; GEMMA, J. N. A modified procedure for stining roots to detect VA mycorrhizas. **Mycological Research**, v. 92, n. 4, p.468-488, 1989.

LACERDA, M. R. B.; PASSOS, M. A. A.; RODRIGUES, J. J. V.; BARRETO, L. P. Physical and chemical characteristics of substrates based on coconut dust and sisal residue for the production of sabia seedlings (*Mimosa Caesalpiniaefolia* Benth.). **Revista Arvore**, v. 30, n. 2, p.163-170, 2006.

LATEF, A. A. Influence of arbuscular mycorrhizal fungi and copper on growth, accumulation of osmolyte, mineral nutrition and antioxidant enzyme activity of pepper (*Capsicum annuum* L.). **Mycorrhiza**, v. 21, n. 6, p. 495-503, 2011.

LEAL JÙNIOR, G.; SILVA, J. A.; CAMPELLO, R. C. B. **Proposta de manejo florestal sustentado do sabià (*Mimosa Caesalpiniaefolia* Benth.)**. 3ª ed. Crato: IBAMA, 1999. 5p. (Technical bulletin).

LEE, S. H.; JI, W. H.; LEE, W. S.; NAMIN, K.; KOH, I. H.; KIM, M. S.; PARK, J.

S. Influence of amendments and aided phytostabilization on metal availability and mobility in Pb/Zn mine tailings. **Journal of Environmental Management**, v.139, p.15-21, 2014.

LINS, C. E. de L.; MAIA, L. C.; CAVALCANTE, U. M. T.; SAMPAIO, E. V. de S. B. Effect of arbuscular mycorrhizal fungi on the growth of *Leucaena Ieucocephala* (LAM.) de wit. seedlings in caatinga soils under the impact of copper mining. **Revista Arvore**, v.31, n.2, p.355-363, 2007.

LIU, H.; YUAN, M.; TAN, S.; YANG, X.; LAN, Z.; JIANG, Q.; YE, Z.; JING, Y. Enhancement of arbuscular mycorrhizal fungus (*Glomus versiforme*) on the growth and Cd uptake by Cd-hyperaccumulator *Solanum nigrum*. **Applied Soil Ecology**, v. 89, p. 44-49, 2015.

LORENZI, H. **Arvores brasileiras: manual de identificaçâo e cultivo de plantas arbóreas nativas do Brasil**. 3.ed. Sao Paulo, 2000. 314p.

LOTFY, S. M.; MOSTAFA, A. Z. Phytoremediation of contaminated soil with cobalt and chromium. **Journal of Geochemical Exploration**, v.144, p. 367-373, 2014.

MA, Y.; DICKINSON, N.M.; WONG, M.H. Beneficial effects of earthworms and arbuscular mycorrhizal fungi on establishment of leguminous trees on Pb/Zn mine tailings. **Soil Biology and Biochemistry**, v.38, n.6, p.1403-1412, 2006.

MAGALHAES, M. O. L.; SOBRINHO, N. M. B. do A.; SANTOS, F. S. dos; MAZUR, N. Potential of two eucalyptus species in the phytostabilization of zinc-contaminated soil. **Revista Ciência Agronômica**, v. 42, n. 3, p. 805-812, 2011. MALAVOLTA, E. **Elementos de nutrição mineral de plantas**. Piracicaba: Ceres, 1980. 251p.

MALAVOLTA, E. **Manual de nutrição mineral de plantas**. Sao Paulo: Ed. Agronômica Ceres, 2006. 638p.

MALAVOLTA, E.; SARRUGE, J. R.; BITTENCOURT, V.C., 1977. Aluminum and manganese toxicity. In: IV Symposium on the Cerrado. Sao Paulo, p. 275-301.

MAN, L. H.; WEN, W. Z.; HONG, Y. Z.; LAM, Y. K.; LING, P. X.; CHUNG, C. K. Interactions between arbuscular mycorrhizae and plants in phytoremediation of metal-contaminated soils: a review. **Pedosphere**, v. 23, n. 5, p. 549 - 563, 2013.

MARSCHENER, H. Role of root growth, arbuscular mycorrhza, and root exudates for the efficiency in nutrient acquisition. **Field Crops Research**, v. 56, p. 203-207, 1998.

MARSCHNER, H. **Mineral nutrition of higher plants**. London: Academic, 1995. 889p.

MARTINS, I. Manganese. In: AZEVEDO, F. A. de; CHOSIN, A. A. da M. **Gerenciamento de toxicidade**. Sao Paulo: Atheneu, 2003.

MCGONIGLE, T. P.; MILLER, M. H.; EVANS, D. G.; FAIRCHILD, G. L.; SWAN, J. A. A new method which gives an objective measure of colonization of roots by vesicular-arbuscular mycorrhizal fungi. **New Phytologist**, v. 115, n. 3, p. 495-501, 1990.

MELO, C. A. D.; GUIMARAES, F. A. R.; GONÇALVES, V. A.; BENEVENUTE, S. da S.; FEREIRA, G. L.; FERREIRA, L. R.; FERREIRA, F. A. Macronutrient accumulation by weeds and corn grown in coexistence in soil with different fertility management. **Semina: Ciências Agrárias**, v. 36, n. 2, p. 669-682, 2015.

MELO, R. F.; DIAS, L. E.; MELLO, J. W. V.; OLIVEIRA, J.A. Potential of four herbaceous forage species for phytoremediation of soil contaminated by arsenic. **Revista Brasileira de Ciência do Solo**, v. 33, n. 2, p. 455-465, 2009.

MENDES FILHO, P. F.; VASCONCELLOS, R. L. F.; PAULA, A. M. de;

CARDOSO, E. J. B. N. Evaluating the potential of forest species under "microbial management" for the restoration of degraded mining areas. **Water Air Soil Pollut**, v. 208, n. 1-4, p. 79-89, 2009.

MORONI, J. S.; SCOTT, B. J.; WRATTEN, N. Differential tolerance of high manganese among rapeseed genotypes. **Plant Soil**, v.253, p.507-519, 2003.

MUKHOPADHYAY, M. J.; SHARMA, A. Manganese in cell metabolism of higher plants. **The Botanical Review**, v. 57, n. 2, 117-149, 1991.

MUKTA, N.; SREEVALLI, Y. Propagation techniques, evaluation and improvement of the biodiesel plant, *Pongamia pinnata* (L.) Pierre - A review. **Industrial Crops and Products**, v.31, n.1, p.1-12, 2010.

NARDIS, B. O. **Growth of five forage grasses in soil contaminated with manganese**. 2012. 15 f. Monograph (Graduation in Agronomy) - Faculty of Agricultural Sciences, Federal University of the Jequitinhonha and Mucuri Valleys, Diamantina. 2012.

NASCIMENTO, C. D. V. do; COSTA, M. C. G.; GARCIA, K. G. V.; SILVA, C. P. da; CUNHA, C. S. de M. Accumulation of nitrogen and micronutrients in legumes submitted to fertilization with organic waste. **Enciclopédia Biosfera**, v.10, n.18, p.109-118, 2014.

NOGUEIRA, M. A.; CARDOSO, E. J. B. N. Microbial interactions in the availability and absorption of manganese by soybeans. **Pesquisa Agropecuària Brasileira**, v. 37, n. 11, p. 1605-1612, 2002.

NOGUEIRA, M. A.; CARDOSO, E. J. B. N. Mycorrhizal effectiveness and manganese toxicity in soybean as affected by soil type and endophyte. **Scientia Agricola**,v. 60, n. 2, p.329-335, 2003.

NOGUEIRA, M. A.; MAGALHAES, G. C.; CARDOSO, E. J. B. N. Manganese toxicity in mycorrhizal and phosphorus-fertilized soybean plants. **Journal of Plant Nutrition**, v. 27, n. 1, p. 141-156, 2004.

NOGUEIRA, M. A.; NEHLS, U.; HAMPP, R.; PORALLA, K.; CARDOSO, E. J. B. N. Mycorrhiza and soil bacteria influence extractable iron and manganese in soil and uptake by soybean. **Plant and Soil**, v. 298, n. 1-2, p. 273-284, 2007.

OLIVEIRA, D. E. C. de; SILVA, A. V. da; ALMEIDA, A. F. de; SIA, E. de F.; JUNIOR, O. R. Response to the inoculation of mycorrhizal fungi and rhizobium in the initial growth of *Acacia mangium* in mining soil in the state of Goiás. **Revista Engenharia na Agricultura**, v. 19, n. 3, p. 219-226, 2011.

OLIVEIRA, J. J. F.; ALIXANDRE, T. F. Biometric parameters of mycorrhized sabià seedlings under phosphorus levels in yellow latosol. **Pesquisa Florestal Brasileira**, v. 33, n. 74, p. 159-167, 2013.

PAVAN, M. A.; BINGHAM, F. T. Metal toxicity in plants. I. Characterization of manganese toxicity in coffee trees. **Pesquisa Agropecuâria Brasileira**, v. 16, n. 6, p. 815-821, 1981.

PAWLOWSKA, T. E.; CHARVAT, I. Heavy-metal stress and developmental patterns of arbuscular mycorrhizal fungi. **Applied and Environmental Microbiology**, v. 70, n. 11, p. 6643-6649, 2004.

PHILLIPS, J. M. & HAYMAN, D. S. Improved procedures for clearing roots and staining parasitic and vesicular-arbuscular mycorrhizal fungi for rapid assessment of infection. **Transactions of the British Mycological Society**, v. 55, n. 1, 158-161, 1970.

PONTES, M. M. C. M.; NETO, T. P. P.; CHAVES, L. de F. de C.; ALBUQUERQUE, S. F. de; OLIVEIRA, J. de P.; FIGUEIREDO, M. do V. B. Double inoculation of β-rhizobium and mycorrhiza in saberzeiro for the recovery of degraded areas. **Pesquisa Agropecuâria Pernambucana**, v. 17, n. ùnico, p. 37-45, 2012.

RABIE, G. H. Contribution of arbuscular mycorrhizal fungus to red kidney and wheat plants tolerance grown in heavy metal-polluted soil. **African Journal of Biotechnology**, v.4, n.4, p.332-345, 2005.

RAMOS, M. H. C. **Removal of color, iron and manganese from water with dissolved organic matter by pre-oxidation with chlorine dioxide, coagulation and filtration**. 2010. 114f. Dissertation (Master's Degree in Environmental Technology) - University of Ribeirao Preto, Ribeirao Preto.

RIBASKI, J.; LIMA, P. C. L.; OLIVEIRA, V. R. de; DRUMOND, M. A. **Sabià (*Mimosa Caesalpiniaefolia*) a tree with multiple uses in Brazil**. Colombo: Ed.

Embrapa Florestas, 2003. 4p. (Technical communication).

RUFINO, G. M. **Bioaccumulation and translocation of manganese in *Sinapis* alba *and Brassica juncea* under increasing doses of this metal in the soil: an alternative for phytoremediation**. 2006. 57 f. Dissertation (Master's Degree in Environmental Sciences) - University of Taubaté, Taubaté.

SANTANA, A. L. **Mineral economy of Brazil 2009 - Manganese**. Brasilia: National Department of Mineral Production/DNPM, 2009. 12p.

SANTIBÂNEZ, C.; VERDUGO, C.; GINOCCHIO, R. Phytostabilization of copper mine tailings with biosolids: Implications for metal uptake and productivity of *Lolium perenne*. **Science of The Total Environment**, v. 395, n. 1, p. 1-10, 2008.

SANTILLÂN, L. F. J.; CONSTANTINO, C. A. L.; RODRIGUEZ, G. A. V.; UBILLA, N. M. C.; HERNÂNDEZ, R. I. B. Manganese accumulation in plants of the mining zone of Hidalgo, Mexico. **Bioresource Technology**, v. 101, n. 15, p. 58365841, 2010.

SANTOS, F. S. dos; MAGALHAES, M. O. L.; MAZUR, N.; SOBRINHO, N. M. B. do A. Chemical amendment and phytostabilization of an industrial residue contaminated with Zn and Cd. **Scientia Agricola**, v. 64, n. 5, p. 506-512, 2007.

SCHNEIDER, J.; OLIVEIRA, L. M. de; GUILHERME, L. R. G.; STÜRMER, S. L.; SOARES, C. R. F. S. Tropical pteridophyte species in association with arbuscular mycorrhizal fungi in soil contaminated with arsenic. **Quimica Nova**, v. 35, n. 4, p.

709-714, 2012.

SCHNEIDER, T.; PERSSON, D. P.; HUSTED, S.; SCHELLENBERG, M.;

GEHRIG, P.; LEE, Y.; MARTINOIA, E.; SCHJOERRING, J. K.; MEYER, S. A proteomics approach to investigate the process of Zn hyperaccumulation in *Noccaea caerulescens* (J & C. Presl) F.K. Meyer. **The Plant Journal**, v.37, p.131-142, 2013.

SCHUβLER, A.; SCHWARZOTT, D.; WALKER, C. A new fungal phylum, the Glomeromycota: phylogeny and evolution. **Mycological Research**, v. 105, n. 12, p. 1413-1421, 2001.

SHUKLA, A.; KUMAR, A.; JHA, A.; AJIT; RAO, D. V. K. N. Phosphorus threshold for arbuscular mycorrhizal colonization of crops and tree seedlings. **Biology and Fertility of Soils**, v. 48, n. 1, p. 109-116, 2012.

SILVA, F. A. S. **Assistat version 7.7 beta**, free distribution. 2013.

SILVA, S. da. **Arbuscular mycorrhizal fungi on the growth, extraction of heavy metals and anatomical characteristics of *Brachiaria decumbens* Stapf. in contaminated soil**. 2006. 82 f. Thesis (Doctorate in Soils and Plant Nutrition) - Federal University of Lavras, Lavras.

SILVA, S. da; SIQUEIRA, J. O.; SOARES, C. R. F. S. Mycorrhizal fungi in the growth and extraction of heavy metals by brachiaria in contaminated soil.

Pesquisa Agropecuària Brasileira, v. 41, n. 12, p. 1749-1757, 2006.

SIQUEIRA, J. O.; POUYÙ, E.; MOREIRA, F. M. S. Arbuscular mycorrhizae in the post-transplant growth of tree seedlings in soil with excess heavy metals. **Revista Brasileira de Ciência do Solo**, v. 23, n. 3, p. 569-580, 1999.

SMITH, S. E.; READ, D. J. **Mycorrhizal Symbiosis**. 2 ed. San Diego: Academic Press, 1997. 605 p.

SOLÎS-DOMÎNGUEZ, F. A.; VALENTiN-VARGAS, A.; CHOROVER, J.; MAIER, R. M. Effect of arbuscular mycorrhizal fungi on plant biomass and the rhizosphere microbial community structure of mesquite grown in acidic lead/zinc mine tailings.

Science of the Total Environment, v. 409, n. 6, p. 1009-1016, 2011.

SOUZA, L. A. de; ANDRADE, S. A. L. de; SOUZA, S. C. R. de; SCHIAVINATO, M. A. Tolerance and phytoremediation potential of *Stizolobium aterrium* associated with the arbuscular mycorrhizal fungus *Glomus etunicatum* in lead-contaminated soil. **Revista Brasileira de Ciência do Solo**, v. 35, n. 4, p. 1441-1451, 2011.

SOUZA, S. C. R.; ANDRADE, S. A. L. de; SOUZA, L. A. de; SCHIAVINATO, M. A. Lead tolerance and phytoremediation potential of Brazilian leguminous tree species at the seedling stage. **Journal of Environmental Management**, v. 110, p.299-307, 2012.

STÜRMER, S. L.; BELLEI, M. M. Composition and seasonal variation of spore populations of arbuscular mycorrhizal fungi in dune soils on the island of Santa Catarina, Brazil. **Canadian Journal of Botany**, v. 72, n. 3, p. 359-363, 1994.

SUGAI, M. A. A.; COLLIER, L. S.; SAGGIN-JUNIOR, O. J. Mycorrhizal inoculation on the growth of angico seedlings in cerrado soil. **Bragantia**, v. 70, n. 2, p. 416-423, 2011.

TAIZ, L.; ZEIGER, E. **Plant physiology**. 3.ed. Porto Alegre: Artmed, 2004. 719p.

TEDESCO, M. J. *et al.* **Analysis of soil, plants and other materials**. 1. ed. Porto Alegre: Federal University of Rio Grande do Sul, 1995.

TURNAU, K.; JURKIEWICZ, A.; LINGUA, G.; BAREA, J. M.; GIANINAZZI-PEARSON, V. 2006 Role of arbuscular mycorrhiza and associated microorganisms in phytoremediation of heavy metal polluted sites. pp. 235-252. *In* : PRASAD, M. N.

V.; SAJWAN, K. S.; NAIDU, R. (eds.). Trace Elements in the Environment. Biogeochemistry, Biotechnology, and Bioremediation. CRC Press. Boca Raton, Florida, USA.

VAN DER LELIE, D.; CORBISIER, P.; DIELS, L.; GILIS, A.; LODEWYCKX, C.; MERGEAV, M.; TAGHAVI, S.; SPELMANS, N.; VANGRONSVELD, J. The role of bacteria in the phytoremediation of heavy metals. In: Phytoremediation of

Contaminated Soil and Water. TERRY, N. and BANUELOS, G., Eds. Lewis Publishers, Boca Raton, FL. p.265-281, 1999.

VASCONCELOS, M. A.; PAGLIUSO, D.; SOTOMAIOR, V. S. Phytoremediation: a proposal for soil decontamination. **Estudos de Biologia: Ambiente e Diversidade**, v.34, n.83, p.261-267, 2012.

VASSILEV, A.; SCHWITZGUEBEL, J. P.; THEWYS, T.; VAN DER LELIE, D.; VANGRONSVELD, J. The use of plants for remediation of metal-contaminated soils. **The Scientific World Journal**, New York, v.4, p.9-34, 2004.

VENDRUSCOLO, D. **Selection of plants for phytoremediation of soil contaminated with copper**.2013. 57 f. Dissertation (Master's Degree in Soil Science) - Federal University of Santa Maria, Santa Maria.

VIDAL, F. W.H. et al. **Rochas e minerais industriais do Estado do Cearâ**. Fortaleza: CETEM/ UECE/ DNPM/ FUNCAP/ SENAI, 2005.

WHO. WORLD HEALTH ORGANIZATION. **Manganese is compounds**. Concise international chemical assessment document n° 12, Geneva, 1999.

WHO. WORLD HEALTH ORGANIZATION. **Manganese**. Geneva. 1981 (Environmental Health Criteria 17).

WILTSHIRE, J. C (2002). Sustainable Development and its Application to Mine Tailings of Deep Sea Minerals. University of Hawaii, Honolulu, U.S.A, 13 p.

WISSEMEIER, A. H.; HORST, W. J. Simplified Methods for Screening Cowpea Cultivars for Management Leaf-Tissue Tolerance. **Crop Science**, v.31, p.435-439, 1991.

XUE, S. G.; CHEN, Y. X.; REEVES, R. D.; BAKER, A. J., LIN, Q.; FERNANDO, D. R. Manganese uptake and accumulation by the hyperaccumulator plant *Phytolacca acinosa* Roxb. (Phytolaccaceae). **Environmental Pollution**, v.131, n. 3, p.393-399, 2004.

ZHANGA, H. H.; TANG, M.; CHENB, H.; ZHENG, C. L.; NIUA, Z. C. Effect of inoculation with AM fungi on lead uptake, translocation and stress alleviation of *Zea mays* L. seedlings planting in soil with increasing lead concentrations. **European Journal of Soil Biology**, v. 46, n. 6, p. 306-311, 2010.

Printed by Books on Demand GmbH, Norderstedt / Germany